Environmental Gov

Climate change is prompting an unprecedented questioning of the fundamental bases upon which society is founded. Businesses claim that technology can save the environment, while politicians champion the role of international environmental agreements to secure global action. Economists suggest that we should pay developing countries not to destroy their forests, while environmentalists question whether we can solve ecological problems with the same thinking that created them. As the process of steering society, governance has a critical role to play in coordinating these disparate voices and securing collective action to achieve a more sustainable future.

Environmental Governance is the only book to discuss the first principles of governance, while also providing a critical overview of the wide ranging theories and approaches that underpin policy and practice today. It places governance within its wider political context to explore how the environment is controlled, manipulated, regulated, and contested by a range of actors and institutions. This book shows how network and market governance have shaped current approaches to environmental issues, while also introducing emerging approaches such as transition management and adaptive governance. In so doing, it highlights the strengths and weaknesses of the different approaches currently in play, and considers their political implications.

This text provides a groundbreaking overview of dominant and emerging approaches of environmental governance, drawing on cutting edge debates and forging critical links between them. The chapters are complemented by case studies, key debates, questions for discussion, and further reading. It is essential reading for students of the environment, politics and sociology, and, indeed, anyone concerned with changing society to secure a more sustainable future.

J. P. Evans is a Senior Lecturer in Environmental Governance in the School of Environment and Development at the University of Manchester. He has an abiding interest in how environmental research underpins urban sustainability, and is currently leading two projects exploring this question in relation to resilience and living laboratories.

Routledge Introductions to Environment Series
Published and forthcoming titles

Environmental Science texts

Atmospheric Processes and Systems
Natural Environmental Change
Environmental Biology
Using Statistics to Understand the Environment
Environmental Physics
Environmental Chemistry
Biodiversity and Conservation, 2nd Edition
Ecosystems, 2nd Edition
Coastal Systems, 2nd Edition

Titles under Series Editor:
David Pepper

Environment and Society texts

Environment and Philosophy
Energy, Society and Environment, 2nd Edition
Gender and Environment
Environment and Business
Environment and Law
Environment and Society
Environmental Policy
Representing the Environment
Sustainable Development
Environment and Social Theory, 2nd Edition
Environmental Values
Environment and Politics, 3rd Edition
Environment and Tourism, 2nd Edition
Environment and the City
Environment, Media and Communication
Environmental Policy, 2nd Edition
Environment and Economy
Environment and Food
Environmental Governance

Environmental Governance

J. P. Evans

LONDON AND NEW YORK

First published 2012
by Routledge
2 Park Square, Milton Park, Abingdon, Oxon OX14 4RN

Simultaneously published in the USA and Canada
by Routledge
711 Third Avenue, New York, NY 10017

Routledge is an imprint of the Taylor & Francis Group, an informa business

British Library Cataloguing in Publication Data
A catalogue record for this book is available from the British Library

Library of Congress Cataloging in Publication Data
Evans, James.
Environmental governance / James Evans. — 1st ed.
p. cm. — (Routledge introductions to environment)
Includes bibliographical references and index.
1. Environmental policy. 2. Environmental management. I. Title.
GE170.E9 2012
338.9'27—dc23 2011021864

ISBN: 978-0-415-58981-9 (hbk)
ISBN: 978-0-415-58982-6 (pbk)
ISBN: 978-0-203-15567-7 (ebk)

Typeset in Times New Roman and Franklin Gothic
by Florence Production Ltd, Stoodleigh, Devon

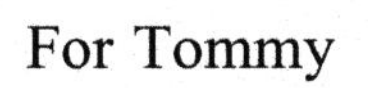

For Tommy

Contents

Illustrations

Figures

Plates

Tables

Analytics of governance boxes

Case study boxes

Key debate boxes

Preface

> It should be possible to explain the laws of physics to a barmaid.
>
> (Einstein, 1879–1955)

This book found its genesis in a rare instance of what could be called "pronoia," or the feeling that life is conspiring to help you. At the start of the 2009 British summer, I was staring out of the window at the rain, wondering how to re-design a module I taught on "theories of environmental governance." It was, I surmised, in a similar shape to the global climate talks. That is to say, dysfunctional in such a way that might be endearing, were it not for the urgency with which it needed to be fixed. Uncomfortable parallels with our atmospheric realities formed in rivulets down the window as I wondered why it was still raining in July. At that precise moment an email arrived from David Pepper, editor of the Routledge Environment series, asking whether I would be interested in writing a book on Environmental Governance. This struck me as a privileged opportunity to produce a text on a topic of pressing importance, and reinvigorate my module along the way.

As the study of how to secure collective action, governance is very much on the environmental agenda. States are wary of making legally binding international commitments to reduce polluting activities, while at the same time direct regulation of industry has fallen from favor. Governance offers a third way. The goal of this book is to provide an introductory overview of the disparate and complex field of study known as environmental governance. Specifically, it aims to introduce the key concepts in environmental governance, drawing together established and emerging work in the field to provide an overview that teases out links, common themes and key challenges.

There is nothing approaching a consensus on either the desirability or the success of governance in addressing environmental issues. In the face of stuttering climate negotiations, Bill McKibben (2007) suggests that we need a new model of governance, while Andrew Jordan and Tim O'Riordan (2003: 223) pithily state, "there ought to be a better way, but nobody has demonstrated what that should be." While anyone concerned with the glacial progress towards a binding climate agreement will be sympathetic to such comments, the proclaimed "Death of Rio Environmentalism" may be somewhat overstated. Nation states are not going to disappear anytime soon, and neither is capitalism or climate change. Governance is here to stay, and assessing its past, present and future is an imperative task for all those seeking action on environmental issues.

Negotiating the field of environmental governance can sometimes feel like hacking through the tangled undergrowth of a jungle with a pocket knife. Governance is like the Medusa of Greek mythology. If it was once a beautiful creature then it has long since become a multi-headed beast, inducing paralysis among those fixed by its gaze. As if it were not enough to be appropriated by multifarious overlapping and competing schools of thought, it is applied to an ever greater set of extant phenomena in the real world. While there are many books with the words "environmental governance" in the title, few attempt a synthesis. This book distils the dominant concepts and emerging forces in environmental governance currently in play.

The material presented here is not intended as a complete account—no introductory text in this field could possibly be—but rather as an exercise in simplification and explication of the key challenges and responses to them. It is not issue-based and makes no attempt to be representative of the range of fields of environmental policy, focusing instead on the key themes and modes of governance currently in play. Given that time is short and the task urgent, I make no apologies for writing this book with Einstein's hypothetical barmaid in mind. Perhaps most importantly, the book endeavors to present a message of hope rather than despair, understanding governance as an opportunity for genuine change.

Acknowledgments

This book could not have been written without the help of many wonderful people, and the opportunity to travel and engage with the inspirational work going on in all kinds of environmental institutions around the world. Special thanks go to Stephanie Pincetl for hosting me in the Institute for Environment and Sustainability at UCLA and providing the most precious commodity of all, space to think. Thanks are also due to the Volkswagen Foundation for hosting me in Germany and inviting me to partake in their "Our Common Future" congress, to Geoffrey Hamilton and Andrey Vasilyev at the United Nations Economic Commission for Europe who shared so much and whose work embodies many of the principles in this book, to Achim Halpaap at UNITAR for engaging, and the wonderful people at UNEP who gave so generously of their time and experience. Thanks also go to colleagues around the world for their support, especially Rob Krueger at Worcester Polytechnic Institute and the anonymous referees who commented on earlier drafts.

The University of Manchester made this project possible by granting me a sabbatical during which the bulk of this book was written, and I would also like to thank my colleagues there, especially in the Geography Department, the Manchester Architecture Research Centre and the Society and Environment Research Group, for their good hearted support and intellectual stimulation. The Royal Geographical Society also deserves a mention for a small grant that allowed me to engage with new fields of research.

Many of the ideas and literatures that have found their way into this book found their genesis in Ph.D. supervisions, and my doctoral students, past and present, deserve thanks for the inspiration that they have provided. A special debt of gratitude goes to all the past students on the M.Sc. Environmental Governance Programme at the University of Manchester, who gamely wrestled through most of the ideas and fields of study discussed in this book (and many, many more besides) with me. Finally, my family and friends deserve special mention for their unwavering support and encouragement. You know who you are, and without you nothing would be possible.

Acronyms and abbreviations

AGGG	Advisory Group on Greenhouse Gases
ANT	actor network theory
CCS	carbon capture and storage
C40	Cities Climate Leadership Group
CDM	Clean Development Mechanism
CEO	chief executive officer
CFC	chloro-fluoro-carbons
CITES	Convention on International Trade in Endangered Species
CO_2e	carbon dioxide equivalent
CoP	Conference of the Parties
CSR	corporate social responsibility
DAD	Decide Announce Defend
DDT	dichlorodiphenyltrichloroethane
EU	European Union
FAO	Food and Agriculture Organization
FSC	Forestry Stewardship Council
G20	Group of 20
GDP	gross domestic product
GEF	Global Environmental Facility
GRI	Global Reporting Initiative
ICSU	International Council for Scientific Unions
ICT	information communications technology
IPCC	Intergovernmental Panel on Climate Change
LULU	locally unwanted land use
MDG	Millennium Development Goals
MIT	Massachusetts Institute of Technology
MUM	Meet Understand Modify
NEPI	New Environmental Policy Instrument
NGO	non-governmental organization
NPM	New Public Management
OECD	Organization for Economic Coordination and Development

QUANGO	quasi non-governmental organization
ppm	parts per million
REDD	reduce emissions from deforestation and degradation
REN21	Renewable Energy Policy Network for the 21st Century
SES	Social–ecological system
ST	socio-technical
UK	United Kingdom
UN	United Nations
UNECE	United Nations Economic Commission for Europe
UNEP	United Nations Environment Programme
UNESCO	United Nations Educational, Scientific and Cultural Organization
UNFCCC	United Nations Framework Convention on Climate Change
USA	United States of America
US EPA	United States Environmental Protection Agency
WCED	World Commission on Environment and Development
WHO	World Health Organization
WMO	World Meteorological Organization
WTO	World Trade Organization

1 Introduction

Intended learning outcomes

At the end of this chapter you will be able to:

- **Appreciate environmental issues as a crisis of governance.**
- **Define the features of governance.**
- **Identify the main challenges and opportunities for environmental governance.**
- **Understand the structure and scope of this book.**

Voltaire's snowflake

> No snowflake in an avalanche ever feels responsible.
>
> (Voltaire, 1694–1778)

Like Voltaire's snowflakes in the avalanche, environmental problems are everyone's fault but nobody's problem. Walt Kelly summed the dilemma up famously on a poster he designed for Earth Day in 1970, saying "We have met the enemy and he is us." This chapter outlines how governance can help address environmental problems, by securing collective action between the diverse groups that make society up, such as businesses, non-governmental organizations (NGOs), government organizations and the public.

The chapter begins by discussing environmental issues as a crisis of governance, or a failure to organize our societies and economies in such a way that they do not harm the environment. As the process of *steering* and enabling collective action, governance has a key role to play in re-organizing society. The chapter then moves on to discuss the implications of uncertainty for those charged with governing the environment, and the opportunities that it presents for change. While the challenges to coordinating action are considerable, there are numerous successful examples from which inspiration can be drawn.

The final section outlines the structure and scope of the book, commenting on its approach, giving an overview of each chapter, and explaining the various boxes and learning tools that are included.

The environment as a crisis of governance

Mike Hulme (2009: 310), a lead author on the Intergovernmental Panel on Climate Change (IPCC) Third Assessment Report in 2001, recently claimed that climate change is a "crisis of governance . . . [not] a crisis of the environment or a failure of the market." Established in 1988, the IPCC gathered vast amounts of evidence to first detect whether the climate was warming, and second to decide whether the warming was attributable to the polluting activities of humans. Following the publication of its fourth assessment in 2007, it is now widely accepted that the answer to both these question is a resounding "yes"—the global climate is warming, and we are to blame.

While the range of scenarios for warming differ in their exact timings, all strongly suggest that a major environmental crisis will occur sometime before the end of the twenty-first century if we continue along our current trajectory of economic development. In other words, "business as usual" will lead us over the edge. The acquiescence of the US administration to enter climate change negotiations in 2009 indicates that this scientific assessment is now widely accepted. How, then, to explain the failure of Copenhagen to secure a binding agreement on emissions reduction? Put another way, if accepted science predicts a forthcoming crisis, then why do we seem unable to act (Zizek 2008)?

A common suggestion is that we do not possess the necessary technology to address the causes of climate change. But a plethora of solutions for polluting industries already exist, ranging from electric cars and wind power through to biodegradable crisp bags and carbon positive housing. The Desertec Foundation, an NGO formed to promote the generation of solar power in deserts, estimates that covering approximately 300 square km of the world's deserts with solar panels

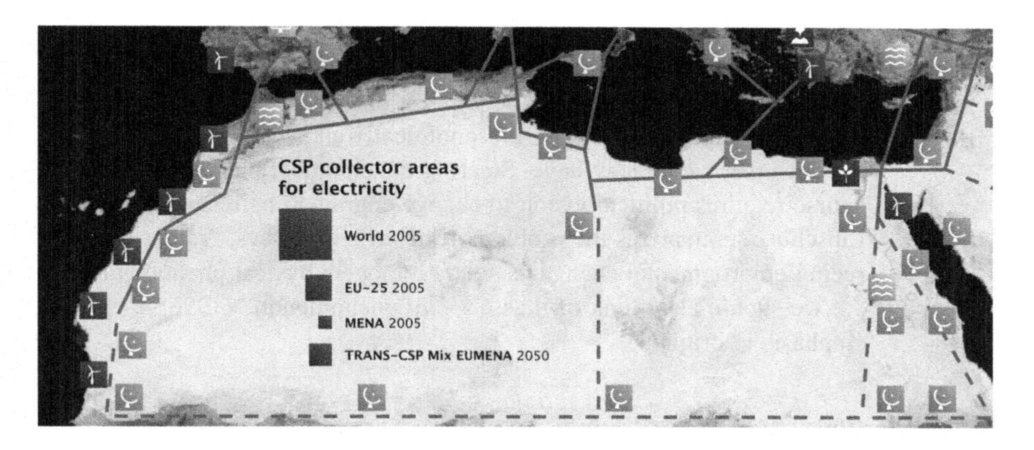

Plate 1.1 ***Area of desert required to supply global energy needs***
Source: reproduced with permission from Desertec Foundation, www.desertec.org.

would produce enough power to supply current global energy needs. Plate 1.1 shows the area of the Northern Sahara required to supply the energy requirements of the world, Europe, and the Middle East nations respectively. The potential is enormous. Why, then, are such technologies not being adopted?

Perhaps the answer is economic. Alternative technologies are notoriously expensive to install and run—certainly more expensive than their existing counterparts. Again, though, this argument falters. Governments around the world subsidize polluting industries such as oil, industrialized agriculture and car manufacturing to the tune of at least 2 trillion dollars every year. These so-called "perverse" subsidies actually work against many stated political priorities. So, for example, subsidizing the price of gasoline prolongs the dependence of the US on foreign suppliers, discourages the development of clean technologies, contributes to traffic congestion (which costs an estimated $100 billion per year), increases carbon emissions and decreases air quality (Myers and Kent 2001). Further, as the 2008 financial crisis showed, there is no shortage of money available to address an emergency that is perceived as urgent.

The answer to these apparent paradoxes is that climate change is no longer primarily a scientific or technological challenge, but a political, social and economic one. The greatest obstacle to mounting solar arrays in Northern Africa is the reluctance of Europe to cooperate with African countries for power. The greatest barrier to implementing new technologies is that we are economically and socially locked-in to the ones that we already have. Steering development onto a different course requires political vision to change engrained beliefs and habits. Lipschutz summarizes the problem neatly when he says, "rather than seeing environmental change as solely a biogeophysical phenomenon . . . we should also think of it as a *social* phenomenon" (1996: 4, emphasis in original).

Defining governance

As the study of how to steer the relations between society and the environment, environmental governance is central to this task. While there is no single school of thought about what governance is, it is generally taken to mean "the purposeful effort to steer, control or manage sectors or facets of society" in certain directions (Kooiman 1993: 2). As Kemp *et al.* (2005: 26) state in relation to the environment, "we cannot assume the wisdom of the market, or any other blind mechanism. Nor can we conjure up the commitment and omniscience required for comprehensively capable central authority. In the establishment of effective governance for sustainability, we must incorporate and also reach beyond the powers of commerce and command—a task best accomplished through understanding, guidance and process." Governance provides a third way between the two poles of market and state, incorporating both into a broader process of steering in order to achieve common goals.

Governance extends the practice of governing to non-state actors, or stakeholders, who have an interest or "stake" in governing, including charities, NGOs, businesses, and the public. Broadening the act of governing in this way brings more resources to bear upon policy problems and maximizes support for decisions. The vast majority of theorists agree that "the role of government in the process of governance is much more contingent" now than before (Pierre and Stoker 2002: 29), shifting from one of rowing to one of steering (Rhodes 1997). While traditional government by the state is a form of governing (Bulkeley and Kern 2006), this book focuses specifically on governance that involves non-state actors (but that may still include the state).

Governance operates by setting common goals or targets, which allow different actors to devise the most suitable ways to reach them. Accordingly, many aspects of governing have been devolved to networks of non-state actors, and new forms of governing have proliferated. Governance is seen by some as the only way to govern an increasingly unruly world, in which the old economic and political coordinates have been eroded by the forces of globalization (Herod *et al.* 1998). To others, the turn to governance undermines the political sphere, replacing democracy with an empty form of proceduralism (Lowndes 2001). This debate extends into the environmental field, and is returned to throughout the book.

The concept of governance emerged from different historical and intellectual lineages, and is used to describe shifts across a number of related but different areas, leading to a degree of confusion concerning the term's usage. In his review, Kooiman (1999) identifies ten different usages of the word:

Governance as the minimal state where governance becomes a term for reducing the extent and form of public intervention, relating to the hollowing out of the state under neoliberalism.

Corporate governance which refers to the way big organizations are directed and controlled, rather than run on a day-to-day basis.

Governance as new public management describing the infiltration of corporate techniques of management and institutional economics into the public sector.

Good governance as a checklist approach to transparent and accountable governing advocated by the World Bank.

Socio-cybernetic governance whereby decisions require the input of multiple actors, all with different knowledges and competencies.

Governance as self-organizing networks in which the state is just one among many actors involved in governing.

Governance as steering as found in the German and Dutch emphasis on the role of governments in steering, controlling and guiding different sectors.

Governance as an emerging international order used by international relations scholars to describe a system of global governance.

Economic governance which focuses specifically on governing the economy or economic sectors.

Governance and governmentality which draws on the French scholar Michel Foucault's analysis of the modern state.

To which could be added:

> *Governance as a form of democratic pluralism* which extends the involvement of the public in decision-making (Kemp *et al.* 2005).

Many of these definitions are returned to and discussed in depth throughout the book. Despite the multitude of contexts in which the word governance is used, and the number of debates surrounding the concept, it captures a very real shift towards more collective approaches to governing societies (Kersbergen and Waarden 2004). A review of the literature finds a good deal of agreement around three core principles of governance: a commitment to collective action to enhance legitimacy and effectiveness, a recognition of the importance of rules to guide interaction, and acknowledgement that new ways of doing things are required that go beyond the state (Kooiman 1999, 2000).

Various modes of governance exist, which facilitate collective action in different ways. Network governance involves voluntary partnerships between diverse stakeholders to build consensus and the collective will and ability to act around a specific issue, while market governance uses financial tools and incentives to steer collective action. Rather than focusing on the specific tools or techniques that are used to address environmental issues, this book focuses on how modes of governance generate different types of collective action and outcomes.

As the practice of governing through cooperation in the absence of a centralized state or dictatorial power, governance has obvious use in addressing environmental problems, which are often global in scope and require a vast range of different people to act collectively. The next two sections outline the challenges of collective action and the opportunities for change presented by environmental issues.

The challenge of collective action

Five key challenges to collective action can be identified in the environmental field. First, scientific uncertainty can make policy-makers hesitant to act. Second, the subjective nature of environmental problems means that solutions can never be right, but merely more or less acceptable to different groups. Third, many environmental problems are transboundary in character, which means that they require international cooperation. Fourth, and closely related to this, the current system of nation states tends to breed competition rather than cooperation. Finally, environmental issues tend to have complex causes that spill across many different areas of human activity, making it hard to coordinate action. It is worth briefly unpacking each of these challenges.

Within the traditional linear model of policy-making, scientists first get the facts right, then decision-makers decide what to do based on these facts (Davoudi 2006, Jasanoff and Wynne 1998). This model appeals to policy-makers because it suggests that there is an objective reality upon which rational decisions can be based. Environmental issues rarely work like this though, because they are characterized by high levels of uncertainty.

Two examples, one simple, and one complex, illustrate the difficulties of establishing scientific facts about the environment. The measurement of a coastline would appear to be fairly straightforward, and yet the answer depends entirely on the scale at which it is measured. Measuring the coast of Canada from geostationary satellite imagery taken from 36,000 km above the earth will overlook smaller inlets. Using accurate maps will produce a larger figure. And if a surveyor walked the entire coast, measuring around every pebble and rock at a certain point in the tidal range, they would conclude that the coast of Canada is infinitely long. Of course, the coastline of Canada does not change length in reality, but the reality we know depends on subjective choices, like how we chose to measure it.

The problem escalates when scientists attempt to understand highly complex systems such as the atmospheric–oceanic system that controls global climate. The fundamental problem is that the climate system has a degree of ontological uncertainty built into it. Ontological uncertainty concerns the actual reality of its functioning, rather than deficiencies in our understanding of it, and no amount of improvement in knowledge or computing power will help. Atmospheric physicists still lack any convincing model of how clouds exchange energy, making attempts to scale up to the entire atmosphere highly problematic (Shackley *et al.* 1998). While the global climate is precisely the system about which politicians want certain knowledge, it is also one of the most chaotic and unpredictable.

But even if scientists could determine the exact adverse environmental effects that might accompany different atmospheric levels of carbon dioxide, they cannot say whether the impact, or risk of the impact, is tolerable. So, for example, if the world continues along a "business as usual" trajectory, then the atmospheric concentration of greenhouse gases is predicted to treble by the end of the century. This gives at least a 50 percent chance that global average temperature increases will exceed 5°C. The general consensus is that extremely bad things like complete ecosystem collapse will happen past 5°C, but again, these are only probabilities (IPCC 2007). This has led scientists to advocate the adoption of a 2°C guardrail, addressed in Key debate 1.1. Ultimately, though, the question of what level of risk is tolerable, and what is "acceptable" in terms of cost and damage, is a political question and the answer will vary depending on who is being asked.

In the absence of scientific certainties, the definition of environmental problems and their solutions will vary according to whose perspective it is seen from, posing what policy analysts call a "wicked problem" (Rittel and Webber 1973). This leaves decision-makers in the unenviable position that their policies can never be right or wrong, but merely more or less acceptable to different groups of people. Climate change certainly seems to belong to this category of problems—people can't even agree whether it is a problem, let alone how to solve it (Auld *et al.* 2007, Levin *et al.* forthcoming). For example, focusing on adaptation rather than mitigation will create huge problems in the future for the developing world, which will bear the brunt of sea-level changes. Mitigating now, however, will lay a greater financial burden on the developed world, creating problems for key sectors of the economy.

The Stern Report on the economic impacts of climate change estimated that the costs of taking strong mitigation measures to prevent dangerous climate change from happening equate to approximately 1 percent of global gross domestic product (GDP) (Stern *et al.* 2006). But there are considerable opportunity costs associated with channeling what equates to some $600 billion into climate change mitigation. This figure represents four times the entire current amount of annual development aid that is sent to poorer countries (OECD 2010), and could be used to alleviate chronic poverty and provide sanitation and education instead of reducing greenhouse gas emissions. A large degree of political inaction on climate change is driven by the fear of making the wrong decision.

This problem is exacerbated by the fact that we have no precursors to learn from. The projected impacts of climate change on the biosphere are substantial and novel, taking humanity into largely uncharted territory (Raudsepp-Hearne *et al.* 2010). While climatic shifts have

Key debate 1.1

The timeframe for averting climate change and the 2°C guardrail

Scientists and policy-makers are increasingly committed to a 2°C guardrail, beyond which the impacts of climate change are considered intolerable. Even if temperatures rise only 2°C above pre-industrial levels scientists estimate that 10 percent of all ecosystems will be transformed, including the loss of 95 percent of the Great Barrier Reef, a 53 percent transformation of tundra ecosystems, 12–26 million people displaced in coastal areas, 1–2.8 billion at increased water stress, and vector-borne disease ranges expanded, among other impacts (Warren 2006). To have at least a 50 percent chance of keeping the global temperature increase below 2°C, atmospheric CO_2 equivalent (CO_2e) concentration will need to be stabilized at about 400 parts per million (ppm). CO_2e concentrations include the effects of other key greenhouse gases, like methane and nitrous oxide, and are generated by calculating the amounts of CO_2 that would cause an equivalent amount warming to each of the other gases. The global atmospheric concentration of CO_2e in August 2010 was 388.15 ppm and rising at 2 ppm per year, so the 400 ppm concentration looks set to be exceeded by about 2017.

Because the climate does not respond immediately to changes in CO_2, it is possible to overshoot the 400 ppm target, as long as the concentration is brought down soon after. Meinshausen (2006) suggests that a peak in CO_2e concentration of 475 ppm, followed by a fall back to 400 ppm over the next century, would still allow the 2°C target to be met. "Peaking" scenarios, where the CO_2 concentration overshoots and then reduces, would require global greenhouse gas emissions to be halved by 2050 relative to 1990 levels in order to limit the global temperature rise to 2°C. A stabilization target of 2°C would require a 60–80 percent cut in emissions from industrialized countries, with a similar abatement path for developing countries in later years. Studies suggest that global emissions must peak somewhere between 2010 and 2020 to achieve this trajectory (O'Neill and Oppenheimer 2002), with steep cuts thereafter. Accordingly, NASA climate expert Jim Hansen (2006) gives the world only a 10-year window in which to avert dangerous climate change.

happened in the past, our ability to adapt in such circumstances is as uncertain as the impacts of climate change itself. It is also difficult to learn lessons from models of governance that have worked in other fields. For example, the United Nations Security Council, which is charged with keeping world peace, deals with specific problems (potential military conflicts), has a common vision (peace), and only needs to include the most powerful countries (those with nuclear weapons). Environmental problems afford no such simplicity. Everyone is implicated in both the problem and its solution, the problem is highly diffuse, it impinges upon many other parts of society, and there is little common agreement as to what outcome is desirable, let alone how to achieve it.

Theories of collective action suggest that rational actors will work together if it makes sense to. So, for example, it could be argued that a rational response to climate change would be for richer countries to make some sacrifice in their current standards of living, in order to help poorer countries adopt cleaner technologies that will avoid massive declines in standards of living for everyone in the future. Unfortunately, history tells us that most actors tend to pursue their own short-term interests.

A game called the "prisoner's dilemma" describes how this situation arises. The prisoner's dilemma involves two (or more) prisoners, who may opt to remain silent or collaborate with their captors to obtain a more lenient punishment. Rational choice theory would dictate that each prisoner should remain silent in order that the captors would only be able to impose a minimal punishment on each (the code of *omerta*, or "silence," by which the Mafia live follows this utterly rational logic). But each prisoner knows that if they remain silent and their accomplice talks then they will receive a very heavy punishment indeed. As a result, both prisoners talk and both receive moderately heavy punishments—the very worse outcome in terms of the amount of punishment suffered overall. The parallel in terms of greenhouse gas emissions is that countries continue to pollute the atmosphere because they cannot be sure that others will stop if they do. Collective action requires trust and frameworks that create certainty for the actors involved.

A closely related problem is that of the "free-rider." Here a group may decide to take action against, say climate change, but it would be rational for an individual country to opt out, as they will accrue the benefits of the collective action without incurring any of the costs. Collective action is also undermined by the asymmetric distribution of costs and benefits, which may lead a state to renege on collective action, or the power of small pressure groups to defeat wider good. For example, companies lobbying for the legalization of genetically modified crops have far more to gain in the short term than the public have to lose from the risks of genetic contamination, even though the overall costs to society in the long term may considerably outweigh the benefits to the companies. Because the interests and thus efforts of the former are highly concentrated, and those of the latter highly dispersed, a pressure group can derail the rational course of action.

Environmental problems often cut across existing political jurisdictions; for example, acid rain is transboundary, while climate change is global. It is not easy to coordinate solutions to these types of problems in a world organized into nation states. A currently fashionable idea is to geo-engineer the atmosphere by releasing particles into the upper layer, which will cause more solar radiation to be reflected back to space and reduce global warming. Studies have suggested that using old military planes could make this measure cost effective (Royal Society 2009), but that there may be a number of major side-effects, ranging from the sky no longer being blue through to a re-opening of the ozone hole. In the absence of a global coordinating body, it is impossible to make decisions concerning these kinds of potential solutions. The global environment occupies a strange position whereby it has become an object of global governance without an accompanying regulatory or legal framework.

Lack of international cooperation has in no small part contributed to environmental problems, as countries have spent the last few centuries competing to gain economic and political advantage over one another by securing and using common resources—a problem discussed in Key debate 1.2. This creates a series of tensions, such as why the developing world should be expected to halt their economic growth when the developed world has already taken the lion's share of resources and emitted massive amounts of pollution. Even if it is accepted that developed countries did not know the ramifications of their polluting activities and cannot thus be held accountable, it will be necessary to convince them to cooperate and reduce their emissions together. And if it is accepted that there is a moral duty for the developed world to assist the developing world, the question becomes how to agree and implement this.

Not only do environmental issues fail to respect national borders, but they result from many different sectors of human activity. Approximately 40 percent of all protein consumed by humans is dependent upon nitrogen fertilizer produced from fossil fuels that create greenhouse gases (Smil 2002). Indeed, national carbon emissions are correlated almost perfectly with national economic output. The only notable decreases in emissions ever achieved in the developed world

Key debate 1.2

The tragedy of the commons

Writing in 1968, Garrett Hardin, an ecologist and trained microbiologist who served as professor of human ecology at the University of California, Santa Barbara, published a paper in the prestigious American journal *Science*, titled "The tragedy of the commons." In it, he put forth the famous argument that environmental problems have no technical solution because they are common resource problems. Using the example of a patch of common grazing land, he argued that it is in the interests of every individual farmer to maximize the number of cattle that they graze on the land, because in the short term each farmer will make more profit. But in the long term, the patch of land will become chronically overgrazed, causing the cattle to die. The resulting destruction of the grazing land affects every farmer, causing a tragedy of the commons.

Almost every environmental problem that we face today can be seen as a tragedy of the commons, and every ensuing failure of nations to cooperate as a playing out of the prisoner's dilemma. Common fish stocks in the ocean have been over-exploited by competing national fishing fleets until they have collapsed. Less tangible resources, like tranquility, have been over-exploited as people insulate themselves from noisy urban environments within ever noisier vehicles. In relation to climate change, the atmosphere has been used by individuals, companies and nations as a global commons in which to dump polluting gases. Invoking philosopher Alfred Whitehead, Hardin claims that the propensity to destroy common resources is a tragedy not in the colloquial sense of an unhappy event, but in the ancient Greek sense of despair at the "remorseless working of things" (1948: 17). The tragedy of the commons occurs not for lack of, but because of, rational actions.

have been the result of economic recession or collapse (for example, the recent financial crisis, or in Eastern Europe after the collapse of the Soviet Union). This relationship works both ways; so when the European Union (EU) and the USA introduced environmental subsidies to encourage farmers to grow biofuels in 2008 they inadvertently caused a world food shortage, as land was turned over to cultivate biofuels. So many aspects of human activity are interrelated with environmental issues that it is exceptionally hard to know where, and at what level, to target actions to address them.

Opportunities for change

Talking about climate change, philosopher James Garvey (2008: 2) notes,

> [c]limatologists can tell us what is happening to the planet and why it is happening, they can even say with some confidence what will happen in the years to come. What we do about all of this, though, depends on what we think is right, what we value, what matters to us. You cannot find that sort of stuff in an ice core. You have to think your way through it.

Climate change conjures up fears and dangers—of losing luxurious lifestyles in the West, or of depriving human needs in the less developed world by hampering development in the name of saving the environment. But it also opens up the possibility of creating a fairer, happier world, and there are plenty of successful examples of collective action from which to draw inspiration.

While scientific uncertainty may have paralyzed political progress on climate change in recent years, there are numerous examples of collective action on environmental issues that have occurred in the absence of certainty. The Montreal Protocol, signed in 1987, was put together by scientists and international organizations, and agreed by major companies and nations, in the absence of absolute scientific proof that CFCs were causing the hole in the ozone layer. The Convention on Biodiversity was signed by 193 states in 1992 despite large levels of uncertainty surrounding rates of extinction, which are reckoned to be somewhere between 74 and 150 per day (Sepkoski 1997). In both these cases, strong alliances between scientists, NGOs and policy-makers created the will to act, even in the absence of incontrovertible evidence. Although much maligned for failing to produce an agreement on emissions reductions, the 2009 Copenhagen Climate Summit did produce the Copenhagen Accord. While not legally binding, this was the first time that all countries had agreed that action was needed on climate change, and represents a major step in international cooperation.

Governance increasingly involves forging transnational networks between businesses, NGOs and other actors that simply bypass reluctant governments. For example, the Forestry Stewardship Certification scheme established in 1993 has certified some 134 million Ha of commercial forests in over 80 countries as sustainable. They have certified the supply chains of corporate giants like Home Depot and

IKEA, and all this has been achieved without any legal regulations in less than 20 years. In extending the practice of governing beyond the state, governance encourages creative responses to the challenges of changing society.

In his book on global innovation, Lessig (2001) argues that new ideas are driven by doubt in the old ones. For example, the Renaissance, arguably the most creative period in Western history, was driven by doubt in the religious coordinates of the old medieval society. The emergence of environmentalism as a major cultural movement in the second half of the twentieth century constituted exactly such doubt in the old industrial society, and prompted many instances of successful change, from the banning of the pesticide DDT through to the United Nations Earth Summits.

Governance is about asking what sort of world we want to inhabit, and how we can coordinate getting there. As Heclo states (1974: 305), "politics finds its sources not only in power but also in uncertainty—men [*sic*] collectively wondering what to do. Governments not only power . . . they also puzzle." Uncertainty and doubt should not be brushed under the carpet, but embraced as creative forces for governance. As the American psychologist William James (1956: 42, quoted in Castree 2010: 185) says, "the world can and has been changed by those for whom the ideal and the real are dynamically contiguous." The current questioning of our oil-dependent society represents a great opportunity to produce an equally creative transition to a low carbon society.

Scope of the book

The goal of this book is to provide an introductory overview of the disparate and complex field of environmental governance. Specifically, it aims to introduce the key concepts in environmental governance, draw together established and emerging work in the field, and provide an overview that teases out links, common themes and key challenges. The book does not try to capture all of the exciting developments in environmental governance that exist across the world, but concentrates on some of the most interesting and influential. Similarly, it makes no attempt to provide representative coverage of the full range of environmental issues (water, biodiversity, pollution and so forth), but rather is thematic, focusing on the key elements of governance.

Governance constitutes a framework for analysis, rather than a theory per se. This distinction is important. A framework indicates what kinds of variables or factors are important, providing an intellectual scaffolding to guide investigation (Schlager 1999). For the most part, environmental governance examines exactly the same things as closely related disciplines like environmental policy, environmental law, environmental management, environmental economics and environmental politics, but through a different lens. As a framework for collective action, governance in its strictest sense concerns the study of institutions, as containers that group different actors together, and rules, which set the parameters within which they interact and act. While there are different modes of governance, such as network governance, market governance, adaptive governance and so forth, they all operate within the framework of governance. The theme of collective action and the institutions and rules that are required to guide it provide common ingredients that link the modes of governance discussed in this book.

By contrast, theories do more than simply identify key factors of interest or importance; they offer an explanation of how the world works, and why things happen the way that they do. As Koontz notes (2003), different theories are appropriate to different circumstances and numerous theories can be brought to bear upon the various factors that make up the framework of governance. For example, institutionalism emphasizes the role of institutions in framing and guiding possible action, while environmental politics focuses on the role and influence of different actors in governing. International relations is a branch of political science that is concerned with the interplay between different nations and other international organizations, while global governance explores the role of civil society in setting international agendas. Geography helps to understand the scales and spaces of governance, while anthropology sheds light on the way in which societies institute rules.

This book draws primarily on the social sciences, based on the premise that fixing environmental problems primarily involves changing the way in which society operates. Different social theories can help understand elements of governance. For example, the social philosophy of Michel Foucault is valuable in understanding how the process of governing relates to the development of the modern state and places what we currently know as governance within its broader historical context. On the other hand, Ulrich Beck's theory of the Risk Society can help us understand the emergence of governance as a response to the uncertainty produced by modern technologies. Each theory explains a social phenomenon, and thus offers a window onto governance.

Structure of the book

The book is composed of 10 chapters, each of which has its own introduction and conclusion that situates it in relation to the key themes outlined above. While the knowledge in earlier chapters is built on in the subsequent ones, each chapter can be used as a stand-alone resource. The book has been structured into two parts. The first part of the book (Chapters 2–4) presents the framework of environmental governance, while the second part (Chapters 5–8) discusses the key approaches, or modes, of environmental governance currently in play. The four modes considered in the second half of the book are not intended to be comprehensive or definitive; there are other modes recognized in the literature, and the activities of governance could have been categorized in other ways. The rationale for selecting these four is to cover the two most influential modes (networks and markets), and two of the more interesting emergent modes that have come to prominence specifically in the environmental field (transition management and adaptive governance). As will become apparent, they are not discrete in practice, and the concept of modes is used primarily as a heuristic device to render the breadth of the subject tractable to analysis. Chapter 9 considers participation and the politics of governance, which cuts across the other four modes. The outline of each chapter is as follows:

Chapter 2 places governance within its broad historical context, tracing how the environment has been governed by nation states, before emerging as an object of global governance. It then explores the causes and consequences of the shift from government (in which the state governs) to governance (in which the state plus non-state actors govern). The main modes of governance (network and market), emerging modes (transition and adaptive) and the theme of participation are presented and the different orders, or levels, of governance are discussed.

Chapter 3 discusses the importance of institutions and rules in enabling collective action, and introduces the key actors involved in environmental governance. Theories of institutionalism are used to help understand the importance of institutional design in shaping collective action, and Elinor Ostrom's work on common pool resources is used to understand how communities develop and enforce their own rules governing resource use. The chapter introduces the key actors involved in environmental governance, including the state, society, business, supra-national organizations, international scientific advisory bodies, NGOs, and sub-national actors.

Chapter 4 addresses environmental governance at the global level, exploring the process of international meetings through which global environmental governance unfolds, and the associated architecture of institutions and rules. It identifies the key conferences, institutions and initiatives that relate to the environment and assesses their legacies. The chapter also discusses the challenge of implementing agreements, and covers key debates surrounding the future of environmental institutions at the global level.

Chapters 5 and 6 cover the main modes of governance that are used to implement environmental agreements. Chapter 5 discusses the network mode of governance, which is characterized by different groups coming together to act voluntarily. The power of networks and their characteristics are outlined, before moving on to consider the importance of transnational networks that operate across and beyond nation states in implementing environmental agreements. The success of certification and auditing networks is discussed, as are the pros and cons of corporate social responsibility in making business more sustainable. The chapter also considers how sub-national actors like cities are forming networks to address climate change. The chapter concludes with an assessment of the strengths and weaknesses of network governance.

Chapter 6 considers perhaps the most influential mode of environmental governance, markets. It begins by outlining the basic principles of the market approach to environmental governance, exploring the examples of the European Emissions Trading Scheme, the Clean Development Mechanism and the Reduced Emissions from Deforestation and Degradation schemes in depth. It also explores different ways in which financial values are placed on the environment, and the implications of doing so for the way in which it is governed. As for network governance, the chapter concludes with an assessment of the strengths and weaknesses of market governance, and it is noted that the state still plays a key role in framing and regulating markets.

Building on this, Chapter 7 addresses an emerging mode of governance known as transition management, which seeks to steer large-scale technological changes in order to make economic growth more sustainable. The concept of transition suggests systemic change, and helps show how climate change mitigation at the level of an entire society might be achieved. The concept of a technological transition is outlined, which depends on niche innovations, and transition management is explored as a mode of governance that encourages experimental innovations. The chapter concludes with an assessment of the strengths and weaknesses of transition management as a distinct approach to governance.

Chapter 8 explores adaptive governance as a mode of environmental governance. Drawing on the ecological concept of resilience, adaptive governance aims to manage social and ecological systems in a holistic way. The core concepts of resilience and the adaptive cycle are outlined, which emphasize continuous change and learning. Adaptive governance holds great appeal as a way to make society more adaptable to climate change, but raises a series of questions for how institutions should be designed. The chapter concludes with an assessment of the strengths and weaknesses of adaptive governance.

Chapter 9 considers the theme of participation and the politics of environmental governance. Participation cuts across the other four modes of governance, as it provides the political vision and values that are required to know in which direction society should be steered. The concepts of risk and the precautionary principle are introduced, and the rationale for involving the public in decision-making is presented. The main models of public participation are briefly outlined, and examples are used to assess the strengths and weaknesses of the participatory model. The chapter ends by considering grass-roots activism and alternative political visions as part of the broader context within which environmental governance takes place.

By way of conclusion, Chapter 10 summarizes the main arguments of the book, reconsidering the evolution of environmental governance and drawing together the discussion of various different modes of governance to compare the ways in which they facilitate collective action. Eight hypotheses on environmental governance are presented, in order to prompt discussion and highlight key areas of future interest.

Text boxes are used throughout the book to provide greater depth and insight into particular topics, focusing on successful and less successful case studies of environmental governance initiatives, key debates and analytics of governance. Case studies have been chosen that shed particular light on a topic, or that are particularly well known in the field. The key debates are intended to take the interested reader into more theoretical depth concerning a particular topic. Finally, the analytics of governance text boxes cover a cutting-edge theory or approach relating to the subject of each chapter, and are intended to be of particular use to those pursuing research in the field.

The end of each chapter lists questions, key readings and web-links that allow important themes to be explored further. A list of acronyms and abbreviations is also provided at the start of the book, although in an attempt to avoid drowning readers in the alphabet soup that characterizes so much of the literature in this field, efforts are made to avoid using them wherever possible.

Parallels and overlaps between the different modes of governance are highlighted throughout the book, and while different approaches are presented in a fairly discrete manner, in reality they are often deployed together to form part of a bigger solution. As the concluding chapter argues, there is no silver bullet, but there are many reasons to be optimistic about governance—after all, it is about changing the world. It is hoped that applying the knowledge in this book will help you to do just this.

Questions

- Do you agree that climate change is now primarily a political problem?
- Are environmental problems distinctive compared to problems from other policy areas?

Key readings

- Hardin, R. (1968) "The tragedy of the commons," *Science*, 163: 1243–48.
- Hulme, M. (2009) *Why We Disagree About Climate Change: Understanding Controversy, Inaction and Opportunity*, Cambridge: Cambridge University Press.
- Meinshausen, M. (2006) "What does a 2°C target mean for greenhouse gas concentrations? A brief analysis based on multi-gas emission pathways and several climate sensitivity uncertainty estimates," in H. Schellnhuber, W. Cramer, N. Nakicenovic,

T. Wigley and G. Yohe (eds) *Avoiding Dangerous Climate Change*, Cambridge: Cambridge University Press, 265–79.

Links

- www.youtube.com/watch?v=EZFkUeleHPY. One-minute cartoon explaining the tragedy of the commons.

2 Governing the environment

Intended learning outcomes

At the end of this chapter you will be able to:

- **Understand the history of governing and the origins of modern government.**
- **Explain the emergence of the environment as something in need of governing.**
- **Appreciate the causes and consequences of the shift from government to governance.**
- **Identify the key modes and orders of environmental governance.**

Introduction

> If the planet has become the content and not the environment, then we can expect to see the next few decades devoted to turning the planet into an art form.
>
> (Marshall McLuhan, 1966)

This chapter considers how the environment has been established as a category or "thing" in need of governing, and the ways in which it has subsequently been governed by an ever-expanding cast of actors. It begins by placing what we now know as governance within its historical context, exploring how national governments traditionally dealt with environmental challenges. It then considers the emergence of global environmental issues, and how these highlighted the shortcomings of traditional regulation at the national level. Piecemeal laws passed to control different types of pollution at the national scale were simply unable to provide the kind of coordinated and strategic response demanded by global environmental problems. These specific pressures were compounded by a more general waning of state power in the face of economic globalization, and an associated right-wing assault on perceived incompetence and waste in the public sector, which together prompted a political shift from government to governance. Within the context of shrinking resources, governments have little choice but to work with other organizations in order to fulfill their duties in many different areas, including the environment.

The second part of the chapter explores the characteristics of governance as a broad concept that refers to the principles, techniques, actors and institutions involved in managing a sphere of human activity. Although many different schools of thought exist concerning what governance is, there is general agreement that it involves sharing the practice of governing with other parts of society, like NGOs, companies and the public. As argued in the introduction, a core goal of governance involves coordinating collective action in order to generate change, but this can be achieved in a number of ways. The chapter outlines the four modes of governing the environment (network, market, transition, and adaptive) that are considered in Chapters 5–8. The concept of orders of governance is also introduced as a way to understand the different levels at which governance can be analyzed.

Governing by government

The idea of government that is familiar to us today, whereby the state has sole responsibility for administering various areas of national policy, only emerged itself in the seventeenth century. Until this time the ruler of a state was responsible for the preservation of the state, rather than with the control and welfare of its population. Government concerned the so-called "high politics" of waging war, making peace, diplomacy, and managing constitutional change. As long as the masses were not actively rebelling they were generally ignored. This all changed in the modern period, as the state began to focus upon "low politics," or administering the needs and everyday affairs of its resident population.

Writing in 1651, Thomas Hobbes described how this shift was based upon the establishment of an implicit social contract between citizens and the state, whereby certain freedoms were forfeited in return for the state providing benefits like law and peace. Having witnessed the

English Civil War at first hand, Hobbes held a fairly pessimistic view of human nature, and believed that the primary purpose of government was to protect society from its own destructive impulses. As well as laying the basis for modern government, the emergence of the social contract prevented unscrupulous monarchs from simply confiscating the property of citizens when they needed it, providing one of the major prerequisites for the unprecedented economic development associated with the Industrial Revolution (North and Weingast 1989).

French social philosopher and historian of ideas Michel Foucault (1977) has argued that the transition from high to low politics was achieved through a transformation in the way that state power was exercised. Rather than using unpleasant forms of corporal punishment to scare the populace into obedience, the modern state began to discipline citizens into certain forms of behavior through institutions like prisons and schools. This idea of discipline characterized the intrusion of the modern state into more and more aspects of the lives of its citizens. Against the nineteenth-century backdrop of rapid industrialization and urbanization, issues like sanitation, food supply, health and nature conservation assumed importance, and, as they did so, new forms of state control emerged to address them. While the City of London passed a measure to control smoke as early as 1273, national environmental protection as we know it emerged in the nineteenth century in response to the problems created by industrialization and urbanization.

With the emergence of modern state administration, "not only does the idea of a measurable and manageable population come into existence, but so also does the notion of the environment as the sum of the physical resources on which populations depend" (Rutherford 1999: 39). The style of governing that characterized the modern nation state was a routine, continuous, and fairly intensive monitoring and regulation of the population and environment, through economic policy, public health, education, sanitation and so on. Foucault coined the term "governmentality" to describe the way in which people internalize the process of governing so that they govern themselves. This idea can also be used to understand how people and the environment are produced as objects of governance, discussed in Analytics of governance 2.1.

Within the modern state, specific branches of expertise like horticulture, modern medicine, civil engineering, and pollution control emerged to administer different problems facing industrial society. This drove a huge expansion of the state apparatus and associated institutions like

Analytics of governance 2.1

Governmentality

The notion of governmentality argues that power is not confined to laws and the state, but is exercised through people and institutions more broadly, with the result that "forms of power beyond the state can often sustain the state more effectively than its own institutions" (Foucault 1980: 73; Foucault 1991). Cultural and political assumptions act to discipline the behavior of people in particular ways. Applying this idea to the environment suggests that problems "are not 'out there' in a pure and unmediated form, but various techniques, procedures and practices construct and produce these fields in such a way that they become both objects for knowledge and targets for regulation" (Bäckstrand 2004: 703, quoted in Rutherford 2007: 294). Governmentality can help understand how environmental principles, techniques, stakeholders and institutions are actively constituted through the practice of governing (Luke 1999, Rutherford 2007). Foucault offers four insights into the exercise of power under the modern state, which can be worked through in relation to the environment (Dean 1999):

Ways of seeing and perceiving. The image of the planet Earth from space represents a classic example of how the technology of space travel, married to the emergence of international environmental NGOs who seized upon it, revolutionized the way in which people saw and perceived the planet. From the solid and seemingly infinite cornucopia that we stand upon, the planet hung in space suddenly looked fragile and finite in ways that it simply never had before (Jasanoff 2004). As discussed in the next section, the Earth viewed from space was a necessary prerequisite for the idea that there was a global environment that was in need of being governed.

Production of regimes of truth which frame how the world is understood. Luke's (1994) study of the WorldWatch Institute argues that their annual report *The State of the World* plays a critical role in establishing the idea that there is such a thing as global resources. Forests and populations (often in the developing world) are intensively monitored, becoming key elements of environmental debate and foci for global efforts. The WorldWatch storyline establishes which things matter in relation to the environment, subsequently framing the actions of networks of NGOs, national monitoring organizations and the various audiences who consume the report.

Technologies and experts. A critical part of this process of framing concerns the ways in which institutions become part of the governing apparatus, promoting new forms of sustainable behavior. For example, various expert organizations have produced tools for living more sustainably, like carbon calculators, and handbooks with tips for reducing domestic energy use.

The formation of bodies and subjects. Foucault offers insights into the way individual subjects can be disciplined to govern or monitor their own behavior, through what he calls technologies of the self. As he states, "individuals are the vehicles of power, not its point of application" (1980: 98). The power we exercise on ourselves has clear links to environmental discourses of self-restraint, like using less energy, consuming less meat, recycling, driving less and so on. As discussed above, experts increasingly bombard us with tools to make ourselves more sustainable—we simply need to apply them to ourselves (Rydin 2007). Governmentality adds depth to understanding of environmental governance by showing how subjects internalize the priorities of environmental experts into their own behavior; a process Agrawal (2005) terms environmentality.

universities, to train experts, house professional and learned bodies and establish techniques for diagnosing and regulating problems.

Expert-led state administration underpinned a "command-and-control" model of governing, which protected common resources by banning or tightly constraining their use. For example, in the USA the federal government expanded and solidified the environment as an object of concern through the introduction of legislation such as the National Environmental Policy Act (1969), the Clean Water Act (1972), and the establishment of the US Environmental Protection Agency (1970) (Landy *et al.* 1994). The command-and-control approach also characterizes global environmental agreements, like the Antarctica (1959) and Moon (1979) treaties that prohibit exploitation of any kind in these places.

Bäckstrand and Lövbrand (2006: 55) summarize the command-and-control approach, saying, "through a detached and powerful view from above . . . nature is approached as a terrestrial infrastructure subject to state protection, management and domination." Laws were made on a piecemeal and largely reactive basis, such that, by the mid-1980s, national environmental policy was a mess of overlapping yet disconnected regulations. The following characteristics were typical (Lowe and Ward 1998):

Low politics. The environment was not high on the political agenda and was not seen as a major concern for central government. Environmental management and regulation was seen as a specialist technical domain outside of the civil service, and was generally pushed away to structures of administration, like agencies and quangos, that were dominated by technical experts and bureaucrats.

Devolved fragmentation. Environmental policy tended to be devolved to local authorities and semi-independent inspectorates, making it hard to act strategically or coordinate priorities between the many different branches.

Disjointed incrementalism. Charles Lindblom (1979), professor of economics and political science at Yale University, coined the term "disjointed incrementalism" to describe the piecemeal and reactive approach to environmental regulation. By this he means that the regulations covering each new environmental problem were simply added to those that already existed, without any attempt being made to identify common problems or address the wider causes of pollution. The long and largely unplanned history of environmental regulation led to a confusing array of institutions and laws, as governments simply muddled through (McCormick 1991).

Within the traditional command-and-control model, national governments dealt with environmental issues as isolated, small-scale technical problems that were fixable with specific laws and procedures. The emergence of global threats from the 1980s onwards, like climate change, acid rain, desertification and biodiversity loss, suddenly and brutally highlighted the shortcomings of this model.

The emergence of the environment as a global problem

Today, the idea that environmental issues are global is taken to be self-evident. But, as with most truths, it began as an idea that had to be nurtured over time. Environmental historian Donald Worster (1977) identifies the start of what he calls "the ecological age" with the test detonation of the atomic bomb in New Mexico in July 1945. For him, this moment more than any other symbolized the fact that humans were capable of inflicting major long-lasting damage upon the planet. Ecology could no longer be delegated to amateur naturalists and university specialists, but needed a permanent place in government. The watershed for the popular environmentalist movement is often taken to be the publication of Rachel Carson's book *Silent Spring* in 1962, which documented the lethal effects of the pesticide DDT accumulating in the food chain (Lytle 2007). As Linda Nash (2006) notes, after *Silent Spring* it was impossible to ignore the fact that humans are a part of, not separate from, ecosystems, and that our actions can and do have grave consequences upon them.

The emergence of environmental science was critical in establishing environmental problems as global issues that required global action to address them. Talking about climate science, Jasanoff and Wynne (1998: 47) argue that its establishment involved "not only the international coordination of assessment and policies but also the difficult task of harmonization at the cognitive level." By "cognitive harmonization" they mean the process by which ways of defining, conceptualizing and measuring objects of research achieve general acceptance among scientists, funders and policy-makers.

Viewing the Earth as a system was fundamental to this process. Emerging from the field of thermodynamics in the 1950s, systems thinking offered a way for scientists to conceptualize the ecological, atmospheric and hydrological components of the planet as part of a single interlinked system of energy exchange. Systems provided a common scientific language for the inter-linkages between species in a food chain eloquently reported by Carson in *Silent Spring*, but also promised a way to measure, predict and manage the performance of nature (Kwa 1987). Set against the backdrop of growing public concern in the 1960s and 1970s, systems thinking became the dominant way in which environmental problems were conceptualized.

Systems thinking provided the conceptual basis for the influential *Limits to Growth* study, carried out by modelers at MIT, which simulated interactions between population, economic activity and resource use in a model called World3 (Meadows 1972). The study showed how over-exploitation of finite resources in a system closed to inputs of energy or matter would lead to cycles of growth and collapse sometime in the twenty-first century. Funded by a high-profile group of businessmen, government leaders and scientists calling themselves the Club of Rome, the study was seized upon by leading figures in the burgeoning environmental movement. The idea of the Earth as a closed system underpinned popular environmental treatises of the time, from Buckminster-Fuller's (1969) language of *Spaceship Earth*, to Commoner's (1971) *Living Machine* and, indeed, Meadows's (1972) own *The Limits to Growth*, which suggested that there are non-negotiable limits to human activity. The idea of limits to growth haunts many of the key ideas underpinning environmental thought today, and is considered in Key debate 2.1.

At around the same time as the *Limits to Growth* report appeared, the US Apollo moon missions were broadcasting pictures of the planet seen

Key debate 2.1

Malthus and the limits to growth

The idea that the environment might set absolute limits to the expansion of human society was first formulated by Thomas Malthus, a British churchman living in the eighteenth century. Observing the deprived conditions of the working classes living in the slums of the new industrial cities, Malthus suggested that humans had over-stepped the limits of natural resources such as fresh air, clean water and food. The reason for misery, he argued, was that while population increases geometrically (1,2,4,8,16 . . .), food supply only increases arithmetically (1,2,3,4,5 . . .) with the result that unchecked population growth will lead to famine and death.

Of course, this projection was not entirely correct. While population has increased almost ten-fold since Malthus' time, the advent of mechanized agricultural production and better yielding crop varieties has allowed food supply to keep pace with population growth. Indeed, a higher percentage of the world's population enjoys clean water today than ever before. This trend has been dubbed "the environmentalist's paradox," as so far the degradation of ecosystems has not led to major adverse impacts on human existence. Nobel laureate economist Amartya Sen (1992) points out that the existence of poverty and hunger has far more to do with the unfair distribution of resources than any absolute limits. Given that the poorest 50 percent of the world adult population own only 1 percent of global wealth, any attempt to blame environmental problems on the procreative tendencies of poor people is, at best, misguided and, at worst, a cynical attempt to shift blame from those who consume most (the rich), to those who consume least (the poor).

Overpopulation is a common target for environmentalists, but the notion that nature presents absolute limits overlooks the fact that resources are defined by human use. For example, oil did not become a resource until the internal combustion engine was invented, and it will cease to be important when it is replaced by alternative fuels.

from space. Seemingly isolated from the inky black nothingness surrounding it, the Earth looked fragile and finite in ways that it simply never did to those whose feet (and viewpoint) had up until that point been firmly planted upon it (Jasanoff 2004). The image provided a perfect visual accompaniment to the rhetoric of spaceship Earth, and has been used widely by environmental organizations to promote a form of globalism. The World Commission for Environment and Development draws heavily on the symbolism of this image, which, in showing no national boundaries or human features, establishes a correspondence between the notions of "one planet" and "one humanity," united in their common home. As the environmental philosopher Sachs (1999) has noted, spaceship Earth produces a dual effect upon the cultural conscience, suggesting that the planet *needs* our care, and that we *can* care for it.

Prompted by an increasing weight of scientific evidence and the vocal lobbying of environmentalists, the United Nations (UN) hosted a series of international conferences on the environment and development from the 1970s onwards. The 1972 Conference on the Human Environment in Stockholm, and the Rio (1992) and Johannesburg (2002) Conferences on Environment and Development, were key events that helped the world to absorb the notion that the global environment was both in need of governing and governable (Biermann 2007).

But the task of governing the global environment clearly could not be addressed by the kind of piecemeal and reactive regulations on which nation states had traditionally relied to manage environmental issues. As Landy and Rubin state (2001), centralized command-and-control works well when it has a few point source polluters to regulate, but breaks down when there are multiple non-point source polluters. For example, it is relatively easy to regulate the emissions from 10 large coal-burning power stations in a single country, but far less easy to monitor the emissions caused by millions of motorists or the effluent discharges from tens of thousands of farms across the world. The taxes and legal regulations typical of the command-and-control approach are ineffective ways to address complex environmental problems. Blanket taxes are too blunt, failing to take account of the different capabilities of organizations to change their behavior, while it is simply too time-consuming and costly to produce specific technical requirements for each and every different industrial sub-sector, and the various operations within each.

The linking of environment and development in the international meetings organized by the UN was anything but accidental. As the Cold War drew to a close in the late 1980s, world leaders were increasingly concerned about environmental security. The old political coordinates of left and right were dissolving, to be replaced by a world that was rapidly globalizing into a single capitalist system. Developing countries harbored serious misgivings about Western environmentalism, fearing that conservation would hamper their economic development. It was in this context that the concept of sustainable development emerged, promising a way to achieve economic development in the developing world while addressing global environmental problems. Defined as "development which meets the needs of the present without compromising the ability of future generations to meet their own needs" (World Commission on Environment and Development 1987: 43), sustainable development allayed the fears of both the developed and developing worlds, uniting them under a banner of environmentally benign capitalist growth. It is to this wider process of economic globalization that we now turn.

Globalization and the hollowing out of the state

Globalization is the process by which national economies around the world have become integrated into a market framework, which allows goods and information to flow across borders. From the 1970s onwards, international organizations like the World Bank and International Monetary Fund aggressively imposed free market policies upon developing countries through a process called structural adjustment, which forced them to pass laws opening their national markets up to international competition in order to qualify for aid and credit. Based primarily upon the ideas of the Chicago School of economics, neoliberalism suggests that the creation of free markets to foster international economic competition is the best way to create prosperity and spread democratic freedom (Friedman 1962). Neoliberals argue that while this process may cause a period of painful adjustment to begin with, as local and national producers are put out of business, it will produce a more competitive and thus successful economy over the longer term.

As popular protests of the sweatshops created by companies like Primark attest, globalization has caused its fair share of problems, and political economists have vociferously attacked neoliberalism, arguing that its policies exacerbate economic inequalities between the rich and the poor (Harvey, D. 2007, Klein 2007). Certainly the collapse of Argentina's economy in 2001 and the disintegration of post-Soviet economies into resource oligarchies in the 1990s raised questions concerning the success of structural adjustment policies at the macro level. At the same time, continuing protectionist measures in developed countries, like subsidizing the prices of agricultural crops in the EU, hint at some level of hypocrisy (or at least selectivity) concerning the developed world's commitment to free market competition. Many of the critiques of various modes of governance that are used in this book are drawn from the field of political economy, discussed in Analytics of governance 2.2.

Margaret Thatcher, the British prime minister who oversaw Britain's wholesale adoption of neoliberal policies in the 1980s, coined the acronym TINA ("There Is No Alternative") to indicate that while the merits of globalization and the free market can be debated, their dominance cannot. In the 1990s, Bill Clinton's campaign team hung the slogan "It's the economy, stupid" on the walls of their HQ, providing a stark indication of the primacy of economic considerations within

Analytics of governance 2.2

Political economy

Political economists study the interaction of economic and political systems, including both the ways in which certain political beliefs can affect the distribution of economic resources, and how economic interests can influence the political activities of governments. The relationship between economy and politics is fundamental to environmental governance; in order to steer society in new directions it is necessary to understand how current political and economic systems support one another (Clapp and Dauvergne 2005). More often than not, political economists provide analyses of how dominant economic interests coincide with political interests to maintain the status quo. A classic example might be the tendency of governments to favor industrial developments, which create jobs and prosperity and are therefore vote winners, at the expense of preventing environmental damage.

Political economists understand governance itself as a symptom of globalization and neoliberalization, and have provided rich commentaries suggesting that governance is simply the latest stage in the political evolution of the global capitalist system. For example, Castree (2008) argues that the demise of the state is just an ideal, and that in reality reforms have required the state to define the nature and extent to which others participate in governing through national laws, the use of monopolies and so on. In this analysis, the state continues to play an instrumental role in supporting the capitalist system by providing new ways for it to exploit nature and papering over the cracks of environmental pollution.

political life by the end of the twentieth century. Economic globalization has led to claims that we live in an "unruly world" which is no longer governable in the traditional way (Herod *et al.* 1998). According to this argument, the old order of sovereign nation states, which divided territories and organized economies, ruled over populations and corporations, disciplined subjects and consolidated identities, is becoming irrelevant, replaced by organs of global governance, like the World Trade Organization, which set rules constraining the actions of national governments. Jessop (1994) suggests that states have been progressively "hollowed out" since World War II, as administrative and political duties have been subsumed by international organizations, and devolved down to regions and localities.

The adoption of neoliberal policies led to the withdrawal of the state from various areas of government as national services like water, gas, and electricity were privatized, and industry and market competition became drivers of change in government. The proliferation of emission trading schemes, carbon offsetting markets and green exchange programs considered in Chapter 6 are part of this wider shift in the political landscape, which has seen state functions devolved to the market.

The shift from government to governance

The period prior to 1990 was an age of "big government," when citizens expected the state to take the lead in providing services, but economic globalization precipitated a crisis of legitimacy in the welfare state. Rather than the state taking sole responsibility for governing, governance provided a way to bring the public, NGOs and business into the process of governing. The hollowing out of the state in terms of decision-making was accompanied by a withering of its capacity for action, making the inclusion of multiple actors in practices of government less of a choice than a necessity in order for the state to fulfill its duties. For example, transport infrastructure like rail and road now requires the participation of private companies both to build and run. National governments simply do not have the human or financial resources to do these things themselves. To some extent, the same things get done in the hollowed out state as in the non-hollowed out state, it is just that they are done by different actors and achieved by different means. As Stoker asserts (1998: 17), "governance is ultimately concerned with creating the conditions for ordered rule and collective

Table 2.1 ***Traditional bureaucracy versus New Public Management***

	Bureaucracy	*New Public Management*
Organization	Hierarchical	Devolved
Procedures	One best way	Flexible
Delivery of services/ goods	Direct government provision	Indirect (e.g. subsidies) or use of non-public agencies
Politics	Administration and politics separate	Need to link in order to ensure accountability
Motivation of workers	Public interest	Can be private as well
Type of activity	Unique challenge	Similar to those faced in the private sector
Personal responsibility	None—tasks merely carried out efficiently	Managers take responsibility for results

Source: adapted from Hughes 2003.

action. The outputs of governance are not therefore different from those of government. It is rather a matter of difference in process."

One of the early ways in which these changes were felt was through the doctrine of New Public Management (NPM), which revolutionized public management and administration. Traditionally, public administration was a bureaucracy concerned purely with enacting policy decisions. The traditional state-led command-and-control approach to governing the environment described in the first section of this chapter was cast in this mold. Political decisions would be taken and the bureaucracy would then administer them. As Table 2.1 shows, bureaucracies were based upon strict procedures and rules that lent them a highly robust and hierarchical character. Describing their emergence in the nineteenth century, the German sociologist Max Weber noted that bureaucracies created a system of authority that was practically indestructible, mechanically efficient, and fair, in so far as it treated people equally.

Originating in the economic crisis of the early 1980s, the New Public Management represented a clear rejection of the bureaucratic paradigm, which was blamed for government inefficiencies and national economic failure in the developed world (Hughes 2003). At the same time, the ideal separation between policy decisions and their administrative implementation was becoming increasingly untenable in the face of a series of perceived failures. New Public Management was driven by thinking from economics and private management that emphasized the need to link performance to rewards. Bureaucracy was suddenly seen as a cumbersome, one-dimensional and largely unaccountable way by which to perform government. Rather than see the challenge of public administration as separate from other operations, it was argued that many of the challenges facing the public and private sector were similar. To be efficient, public administration should adopt more flexible procedures and performance management from industry.

To ensure value-for-money, public services were either privatized entirely or redesigned to operate in line with market principles, in real markets if they existed, or in new "pseudo-markets" if they did not (Bailey 1993). The introduction of managers in hospitals who are rewarded if they meet a set of predetermined targets for criteria like service and patient satisfaction is an example of a pseudo-market. New Public Management transmitted the broader changes associated with globalization and neoliberalism to the public sector, prompting a paradigm shift in the ways that public policy was implemented. In the environmental sector, command-and-control approaches were replaced with so-called "New" Environmental Policy Instruments (NEPIs), like environmental taxes, voluntary agreements, eco-labels, and tradable permits, which required the participation of many actors beyond governments (Jordan *et al.* 2003). The application of these kinds of instruments in the environmental field is discussed in depth in Chapters 5 and 6.

Modes of governance

Governance seeks to coordinate collective action between actors, but there are a number of different ways in which this can be done. Modes comprise bundles of rules that guide interaction based on general principles about how actors are best motivated. Three different modes of coordination are generally recognized in the literature: hierarchy, network, and market.

Hierarchy is the mode of governance that most resembles traditional government, whereby there is a clear pyramid of control through which decisions taken at the top are subsequently passed down to those below. As Table 2.2 shows, stakeholders are tied to each other formally as employees, and are bound by the authority of the organization to perform their duties. This mode of governance is very rigid, adhering to routines and simply enacting decisions that are made higher up. The benefits of this mode of governance are that it establishes a clear route to a desired outcome, is durable and stakeholders are committed to the organization; but it tends to breed a lack of innovation and inflexibility. The organization of most private companies or public administrations adheres to this model. This book does not devote a separate chapter to hierarchy as a mode of governance, as it does not act by coercion or steering, but by force. As such it is better seen as a mode of *governing*, and is discussed earlier in this chapter in relation to the traditional command-and-control approach to the environment.

Network governance is the mode most commonly associated with the concept of governance (Rydin (2010) calls it pure governance), whereby autonomous (separately empowered) stakeholders work together to achieve common goals. The concept of the network captures the expanding range of people involved in governance, emphasizing the connections between them as independent actors rather than their organization within an overall hierarchy. As Table 2.2 demonstrates, stakeholders are bound together by the belief that they have complementary strengths that will allow them to achieve shared goals more effectively if they collaborate. Working together provides mutual benefits by creating shared agendas for action, and pooling

Table 2.2 ***Hierarchy compared to network and market modes of governance***

	Hierarchy	*Network*	*Market*
Basis of relationship between members	Authority	Complementary strengths and trust	Contract/property rights
Means of interaction	Routines	Relational	Prices
Tools for governing	Regulation	Collaboration	Financial incentives
Approach to resolution	Administrative	Reciprocity	Bargaining
Flexibility	Low	Medium	High
Commitment of members	High	Medium	Low
Ethos	Formal	Mutual benefits	Suspicion
Choices made by members	Dependent	Interdependent	Independent
Role of the state	Laws, rules and regulations (the "stick")	Encourage voluntary behavior (persuasion)	Economic incentives (the "carrot")

Source: adapted from Powell 1991, Lowndes and Skelcher 1998 and Rydin, 2010.

resources to enable them to do things they would not otherwise be able to.

The means of communication and approach to conflict resolution depend heavily on the levels of trust the stakeholders have in each other. Networks are more flexible than hierarchies, as they do not require formal employment contracts, and can thus be more responsive to emerging needs and opportunities. The disadvantage is that there are few formal constraints preventing stakeholders from leaving the network, making them less robust.

Market modes of governance bind stakeholders together as suppliers and consumers of particular resources or products. The creation of property rights and contracts between stakeholders allows them to trade resources between one another according to the laws of supply and demand. Price provides the means of communication between stakeholders, who are largely free to enter and exit the market according to their own volition. Financial incentives provide motivation for action, and can be used to enhance the power of certain stakeholders over others. The flexibility of this mode is offset by the lack of commitment of its members, who may be motivated purely by profit, rather than any belief in the overall purpose of the political process.

Network and market modes of governance require specific types of institutions and rules, which privilege certain stakeholders over others. So, for example, the network mode favors the creation of umbrella organizations and NGOs as network facilitators, while the market mode emphasizes the role of private companies. Specific modes also cast stakeholders in different roles. So, the network mode casts the public as environmental citizens, motivated by common ethical concerns, while the market mode casts the public as consumers, motivated by financial incentives. Similarly, the kinds of institutions required by each will vary according to the role of the state. Institutions designed to create laws and enforce regulations (as required by the market mode) will require very different resources and competencies to those that are required to encourage and support voluntary behavior (as required by the network mode).

While the ideal types of markets and networks serve as a useful starting point, in the real world "price, authority and trust are combined with each other in assorted ways" (Bradach and Eccles 1991: 289), and hierarchy, network and market governance are all effective at addressing environmental issues in differing contexts (Steward 2008). For example, in relation to the automobile industry, simply banning the use of leaded petrol through a top-down directive has been an incredibly effective way to reduce pollution from automobiles. In 2008, the former chairman of Shell, Sir Mark Moody-Stuart, advocated legal restrictions on fuel efficiency in new cars to encourage innovation and improve standards. Elsewhere, voluntary measures have achieved substantial improvements in sustainability performance. For example, Japan's Top Runner program identified the leading performer (the top runner) for sustainability in the automotive industry, and developed a timetable with other car producers for this to become the standard. Part of the success of the Top Runner program was achieved through publicizing the environmental performance of different producers, which opened them up to public scrutiny and stimulated innovation as they sought to outdo each other (Nordquist 2006).

Increasingly, financial incentives are being used to encourage more sustainable technologies, like the "feed-in tariff" for renewable energy in Germany, which subsidizes producers of renewable energy by guaranteeing a price for energy sold back to the main grid (Carbon Trust 2006). Subsidies for the purchase of electric cars have also been made available in some countries in order to influence the economic behavior of car producers and the people who buy them. Far from making them antithetical to one another, the differing characteristics of each mode of governance makes them complementary, and a critical challenge concerns how to ensure that the appropriate blend of governance is used.

In addition to the network and market modes of governance, two supplementary modes are addressed in this book: transition management and adaptive governance. These are less well established modes of governance in the literature, but build upon the market and network modes respectively and are gaining influence specifically in the environmental field.

Transition management seeks to steer large-scale technological change in a sustainable direction, by creating conducive economic and political conditions, known as niches, for innovations to develop and subsequently spread through society. As Table 2.3 shows, stakeholders share a common interest in innovation, and the main means of interaction is through "evolutionary pressures," whereby wider political and economic forces select some innovations to succeed and others to fail. Like markets, stakeholders are largely free to participate according to their own volition, but unlike markets this is a purely managerial

Table 2.3 ***Two emerging modes of environmental governance***

	Transition	*Adaptive*
Basis of relationship between members	Innovation	Complementary knowledge and resources
Means of interaction	Evolutionary	Learning
Tools for governing	Niche management	Monitoring and experimentation
Approach to resolution	Political	Reciprocity
Flexibility	Medium	High
Commitment of members	Low	High
Ethos	Managerial	Mutual benefits
Choices made by members	Interdependent	Interdependent
Role of the state	Economic and policy incentives	Encourage

mode of governance that provides incentives to steer innovation. This mode of governance is moderately flexible—the state can change the economic and political incentives, but they require time to take effect.

Adaptive governance brings actors with a stake in a social–ecological system, for example a fishery, together in order to monitor that system and change their behavior accordingly. This mode extends network governance to include ecological systems, with stakeholders bound together by the belief that they have complementary interests which will allow them to manage a resource more effectively if they work together. Governing takes place through a process of monitoring and experimentation that facilitates iterative learning and adaptation in the context of a changing environment. Success depends on the levels of trust the stakeholders have in each other, as adaptive governance requires stakeholders to be willing to learn from one another. As Table 2.3 shows, the entire rationale of this mode of governance is to be highly flexible, allowing for change and adaptation.

As for network and market modes, transition management and adaptive governance privilege specific actors, and cast them in different roles. For example, transition management emphasizes high-level collaboration between policy-makers and private business, while adaptive governance emphasizes community knowledge and learning. The relative strengths and weaknesses of networks, markets, transition and adaptive modes of governance are returned to in the final chapter.

Orders of governance

While modes refer to different types of governance, analysts also distinguish between the different levels at which governance occurs, referred to as first-, second- and meta-governance orders (Kooiman 2000). As Figure 2.1 shows, first-order governance covers the way that problems are dealt with directly through action and implementation. In relation to climate change, for example, first-order governance might involve deciding on the mix and proportion of renewable energy in an overall national energy policy. The governance challenge at this level involves devising a decision-making process that is legitimate (includes the people who will be affected by the decision) and efficient (includes the best knowledge and expertise on the subject).

Second-order governance is concerned with the context in which the first order takes place, focusing on institutional design and the creation of policy instruments and programs to steer first-order governance. Taking the example of climate change once more, a classic second-order governance challenge facing governments concerns how to institutionalize climate change in order to make effective and fair

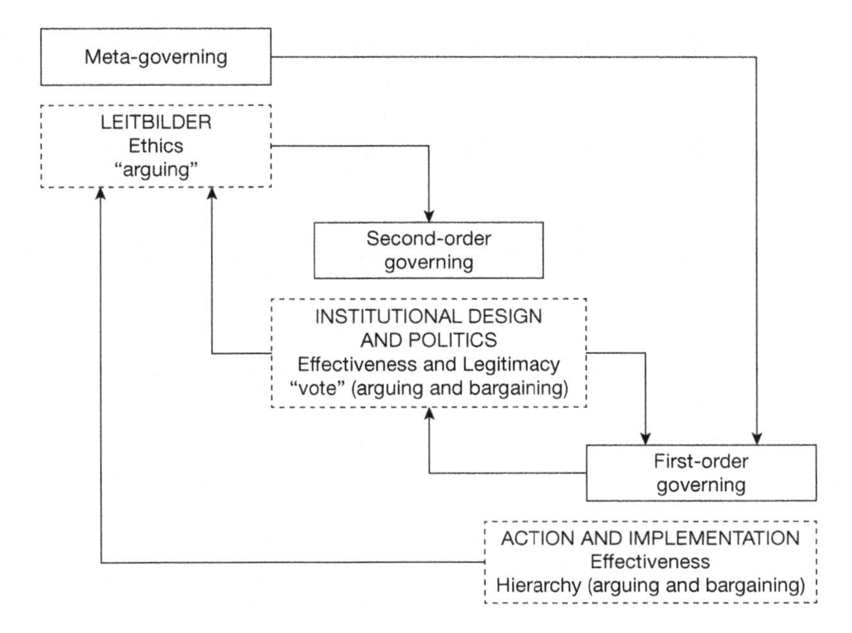

Figure 2.1 ***The three governing orders***

Source: adapted from Heinelt 2007.

decisions. The environmental governance literature often implicitly focuses on second-order governance, and the importance of institutions is unpacked in more detail in the next chapter.

Meta-governance is concerned with the governance of governance. The term *Leitbilder*, used in Figure 2.1, means guiding principles, and meta-governance tends to unfold through ethical arguments and debates concerning the norms within which problems are framed. According to Bob Jessop (2003), meta-governance concerns the organization of the conditions for governance, or how the contextual factors shape the way in which institutions are built and problems are presented. Again returning to climate change, cultural interventions like Al Gore's film *An Inconvenient Truth* have played a role in establishing a shared understanding of (or at least a shared debate over) climate change as a problem, including what it is, why it is occurring, and what needs to be done about it. Meta-governing opens the field of governance up to a far broader set of considerations, including the role of the media (see Key debate 2.2) in shaping public opinion.

As Figure 2.1 shows, the three orders correlate to different levels at which governance can be used to approach a problem. One of the key insights of this framework is to explain why measures that are valid and useful at one level may not necessarily be valid at another. For example, it is probably futile to try to roll out feed-in tariffs (first order governance) for renewable energy in countries without institutions in place that are capable of coordinating energy suppliers and government departments (second order governance), or before promoting some wider understanding among the population of what renewable energy is and why it matters (meta-governance).

Discourse is a key concept that is used to understand how environmental issues can be framed in different ways, and what the implications of doing so are. Discourse literally means "connected utterances," and concerns the way in which communication normalizes certain meanings in relation to specific subjects. Discourses are unavoidably political, in that normalizing some meanings and actions simultaneously excludes others (Fairclough 1992), and represent a powerful way to explore environmental issues, which are by definition uncertain and thus open to multiple interpretations. For example, Thomas and Middleton (1994) have shown that the discourse of desertification in the 1980s owed more to cultural assumptions concerning the advancing deserts of North Africa that were set by colonial explorers than to empirical

Key debate 2.2

The role of the media

While the media are not strictly a part of governance, their influence over public opinion and the communication of environmental science means that they play an important role in meta-governance. As Bennett states (2002: 10), "few things are as much a part of our lives as the news," and the media have great potential to generate legitimacy for collective action. On the one hand, the media reinvigorate the social sphere by stimulating public debate (Morley and Robins 1995), but on the other, research has highlighted the tendency of the media to distort environmental issues. Max Boykoff's (2007) work has shown how the climate debate is presented in the US media as being highly contentious, despite increasing scientific consensus, in order to generate more interest around the story. As Dan Brockington (2009) has concluded in his study of celebrity and conservation, the media are primarily concerned with entertainment, not information delivery.

Examining 600 newspaper articles and 90 TV and radio reports on environmental issues, Ereaut and Segnit (2006) concluded that the dominant message conveyed by the media was alarmist, focusing on the potentially disastrous effects of climate change. Narratives of doom make good headlines, but make people feel less able to take positive action. The problem here is that exciting events are necessary to maintain media interest in the environment. As Downs's (1972: 39) attention cycle, which describes the "systematic cycle of heightened public interest and then increasing boredom with major issues," argues, events soon become old news if they have few exciting events associated with them.

measurements. In framing this issue in a certain way, the discourse of desertification exerted a powerful influence over how it was addressed, supporting high levels of funding for research into the phenomenon, and driving land management policies that excluded local herders, who were seen as part of the problem. The way in which environmental issues are framed has important implications for how they are addressed, and by whom. Case study 2.1 discusses a discursive framing that has gained political influence recently—the idea that climate change presents a threat like terrorism.

Chapter 9 explores how wider sets of stakeholders can participate in the different orders of governance, by involving actors who will be affected by a decision in making it. Participation generates legitimacy, and improves collective action by bringing the knowledge of different actors to bear upon the decision-making process. Participation usually takes

Case study 2.1

Framing climate change as a security threat

In 2004 the UK government's chief scientist, Sir David King, claimed that climate change presents a greater threat to the world than terrorism, questioning the disparity between the huge quantity of resources being poured into the war on terror and the paltry amounts being committed to combat climate change. These priorities have certainly not changed—the $3 billion pledged by the USA to the rapid response climate fund at Copenhagen would have covered the cost of roughly 72 hours of the Iraq/Afghan war that was being waged in 2010. In the same year, Peter Schwartz, CIA consultant and former head of planning at Royal Dutch/Shell Group, and Doug Randall, of the California-based Global Business Network Climate Change, wrote a report for the Pentagon in which they argued that climate change "should be elevated beyond a scientific debate to a US national security concern."

Framing climate change as a security threat was an attempt to move it up the political agenda. Like terrorism, climate change carries with it suffering and death as possible outcomes, so if terrorism scares people, then climate change should too. But framing climate change as a national security threat changes the discourse in subtle ways. For example, it suggests that it is something external to society, which must be fought by individual nation states, rather than a problem that is "in here" and which requires collective action between states. Such a framing has also generated some unlikely alliances. In the USA the Iraqi war prompted American environmental groups to lobby potential 4×4 drivers on the ground that by consuming more gas they were lining the pockets of Middle Eastern countries associated with Islamic terrorism. Right-wing conservatives and environmentalists were briefly united around the idea that more fuel efficient vehicles were desirable, as a patriotic choice for the country, and as an ecological choice for the planet, respectively.

place through a formal process of dialogue, adhering to procedures and feeding outcomes into decisions. For example, in relation to renewable energy, public participation can improve first-order governance by identifying the initiatives that will be most likely adopted by people. Participation can also enhance second-order governance, by contributing to institutional design, and even clarifying the levels of trust that the public have in different institutions. At the meta-governance level, participation can reveal wider cultural preferences and help establish overarching political visions to steer governance. The theme of participation cuts across the three orders of governance.

Conclusions

In outlining the historical context from which environmental governance emerged, it becomes clear that governance is not simply the "next best approach" to environmental issues, but that it has a history and a context that led policy-makers to it. The transition from government to governance has been gradual, evolving and constantly changing, with no set blueprint. Evidence concerning the implementation of New Public Management reforms in the developed world suggests that the process has been far more piecemeal than some of the literature suggests, and highly dependent upon what has gone in the past.

While the shift from government to governance is certainly not restricted to the environmental sphere (neoliberalism, New Public Management, crisis of legitimacy, and public demands for improved services have driven change in all sectors), it has been pronounced by a number of specific challenges that the environment presents. The history of environmental policy evolution, environmental problems, the way they are perceived, and the policy instruments we have to act upon them in different places and times, reflects a need for new frameworks that can incorporate broader sets of actors and more flexible approaches in order address the problems of global environmental change. The different modes of governance and the orders at which they operate provide a rich array of resources to address environmental issues. Before moving on to consider these modes in practice, it is necessary to understand who the key actors are in environmental governance, and how they are grouped and guided by institutions and rules.

Questions

- To what degree has "cognitive harmonization" been achieved in the environmental sphere?
- Examine how an environmental issue of your choice is governed at the first, second and meta-governance levels.

Key readings

- Jasanoff, S. (2004) "Heaven and Earth: images and models of environmental change," in S. Jasanoff and M. Martello (eds) *Earthly Politics: Local and Global in Environmental Governance*, Cambridge, MA: MIT Press, 31–52.

- Jordan, A., Wurzel, R. and Zito, A. (2003) "Comparative conclusions. 'New' environmental policy instruments: an evolution or a revolution in environmental policy?" *Environmental Politics*, 12: 201–24.
- Stoker G. (1998) "Governance as theory: five propositions," *International Social Science Journal*, 50: 17–28.

Links

- www.radford.edu/~wkovarik/envhist/. Home of the Environmental History Timeline, in its own words "an independent project by an American scholar, not funded by any government agency or supported by any foundation or advertising."
- www.aip.org/history/climate/index.htm. The Discovery of Global Warming. Spencer Weart's hypertext history of how scientists came to understand how people are causing climate change.

3 Institutions, rules, and actors

Intended learning outcomes

At the end of this chapter you will be able to:

- **Understand the importance of institutions.**
- **Appreciate how rules govern action.**
- **Identify the key actors involved in environmental governance.**

Introduction

As a framework, governance places great importance on the role of institutions in grouping different actors together, and rules in steering their action. This chapter considers the role of institutions and rules in enabling collective action, before moving on to identify the key actors in environmental governance. In doing so, it grounds the discussion in subsequent chapters.

Drawing on theories of institutionalism, the first section explores what institutions are, how they function, and the importance of designing them appropriately. The work of Elinor Ostrom, who became the first female Nobel Prize winner for economics in 2009, on common pool resource management is used to categorize the different types of rules that shape action, paying particular attention to those that enable the sustainable governance of resources.

The second part of the chapter identifies the key actors involved in environmental governance, including states, society, business, supra-national organizations, sub-national bodies, international scientific advisory bodies and NGOs. Debates concerning the exact status of national governments under conditions of governance are discussed, as are the roles of society and business in addressing environmental issues. Particular attention is paid to the role of the United Nations, specifically the United Nations Environment Programme, the origins and role of NGOs, and the Intergovernmental Panel on Climate Change, as key actors in global environmental governance.

Institutions

The question of what constitutes an effective institution has occupied political economists since the nineteenth century. According to the dictionary, an institution may be an "established law, custom or practice," and as such they concentrate "the traditions and conventions which evolve in a free society" (Hayek 1948: 23, quoted in Shogren 1998: 255). The shift to governance has been accompanied by a proliferation of institutions, as functions previously performed exclusively by governments have been devolved to actors working either separately or at arm's length from the state. In addition to generating the need for institutions to group and coordinate non-state actors, governance places great importance upon institutions as arbiters between the interests of different stakeholders. As Rydin (2010: 96–97) states, "institutions bind actors together into arrangements and patterns of behavior that exhibit strong path dependencies . . . Actors learn to behave in accordance with institutional norms and this reinforces certain behavior." Institutions are not simply political or administrative units, but guide collective action by setting the "rules, norms and practices, which structure areas of social endeavor" (Coaffee and Healey 2003: 1982).

Public policy scholar Vivien Lowndes (1996: 182) suggests that institutions display three defining characteristics:

Institutions operate at the meso-level. Institutions link the broader social fabric to the day-to-day decisions and actions of individuals, being created and shaped by individuals, but structuring what they can subsequently do. Institutions simultaneously open up new fields of action, but constrain the form which that action can take. For example, the establishment of the Environmental Protection Agency (US EPA) in 1970 effectively created the environment as an object of governance in the USA. Making the environment an issue of federal concern meant that environmental policy-makers suddenly had far more powerful regulatory actions available to them than they had before, but the way in which the Environmental Protection Agency was structured meant that they were largely limited to pollution control.

Institutions have formal and informal aspects. Institutions operate through sets of codified rules, but they are also characterized by habitual actions and traditions that guide behavior in relation to governing. Rules can be implicit and informal and yet still order the way in which things are done. For example, many alpine herding communities practice transhumance, moving their livestock to higher altitudes over the warmer summer months to allow pasture at lower altitudes to recover for the winter. The practice of transhumance is not codified in a set of rules—herders are not instructed to do it—but rather the seasonal movement of livestock is built into the culture and calendar of alpine herding communities like the Swiss Appenzeller, with festivals marking seasonal movements, and practices like yodeling reflecting the need to communicate at high altitudes. Institutionalism views informal traditions and habits to be as important as formal rules in the study of institutions.

Institutions generate more legitimate decisions and are stable over time. The actions of institutions are seen as more legitimate than those of individual actors, because they are generated by multiple actors, in accordance with set rules, and are relatively stable over time. Some religious and educational institutions, like the Vatican Church in Rome or Oxford and Cambridge universities in the UK, have existed for over 500 years with very few changes to their basic institutional structure. Instituting a decision-making process makes it more transparent and accountable than simply making decisions either individually or behind closed doors.

The role of institutions in shaping how political decisions are taken and enacted was highlighted in a paper written by James March, an American behavioral psychologist, and Johan Olsen, a Norwegian political scientist, in 1984. Previously, government decision-making was interpreted as a consequence of individual behavior, so the decisions of bureaucracies could be attributed to the attempts of the individuals working within them to achieve their own ends (often based upon striking a balance between securing personal promotion and achieving the goals of the organization). Coining the term "new institutionalism," March and Olsen suggested that, while the influence of individual behavior is important, decisions are shaped to a large degree by the pre-established rules and procedures through which institutions respond to real world issues.

The insights of institutionalism have achieved widespread acceptance, creating a number of implications for governance (Pierson and Skocpol

2002). The new institutionalism emphasizes that institutions are not static things, but dynamic entities that require constant maintenance and reproduction through sets of procedures and rules that become habitual (Lowndes 1996). As Bevir and Rhodes (1999: 225) state, an institution "is created, sustained or modified through the ideas and actions of individuals," embedding a specific set of dilemmas, beliefs and traditions. Speaking about the governance of natural resources, Bridge and Jonas (2002: 760) note that institutions are often established as the result of historical struggles, manifesting their outcomes and bringing them to bear upon the way in which current decisions are made; "by defining what is economically, technologically, and politically possible at particular moments, such institutions can lend coherence and stability to efforts to extract, process, market, and consume natural resources." The weight of history lends institutions path dependency, whereby, once instituted, a decision-making process tends to produce similar decisions over time. Given their influence over decision-making, the way in which institutions are designed is critical.

The question of how to manage institutional change is significant in the realm of environmental governance, which is characterized by rapidly evolving problems. Institutions can become suboptimal due to ossification, over-complexity and the predominance of self-interest, but radically restructuring institutions too often reduces their capacity to act and erodes public legitimacy (Jones and Evans 2008). For example, climate change is driving a restructuring of institutions to reflect the greater importance of energy in the environmental field, but there are numerous ways in which this can be done. Case study 3.1 discusses the different ways in which countries have institutionalized climate change within their governments, highlighting the strengths and weaknesses that accompany each institutional solution.

Rules

By its very definition, governance allows more people to participate in governing, raising important questions concerning who is allowed to participate and how. Rules are critical to securing cooperation, as they provide certainty and security for different actors. They constitute the key variable that can be manipulated in governance, determining who gets to govern, and what they are allowed to do when they get there. As the cartoon in Figure 3.2 shows, setting overarching rules prescribing set courses of action tends to fail, as they cannot capture the diverse

Case study 3.1

National institutions and climate change

Governments throughout the world are shuffling their institutions in the face of climate change, but there is no single blueprint concerning the best way to institutionalize the problem (McIlgorm *et al.* 2010). A stratified sample of 31 countries with different political systems, taken from the developed, post-socialist and the developing worlds, reveals five broad ways in which countries are institutionalizing climate change (Chisholm *et al.* 2010):

Housing climate change within a broad institution that has not been recently merged. This is one of the least focused approaches to national climate change governance, which simply incorporates it into an existing area of government activity. Examples include Canada, Japan, New Zealand, South Korea, Sweden, and Venezuela.

Creating a specific institution devoted to climate change. Usually formed through the separation, amalgamation or creation of departments/ministries, these kinds of institutions often report directly to the president, and are relatively independent. Examples include Australia, Brazil, Chile, China, Denmark, Indonesia, the Maldives, Panama, South Africa and the UK.

Fragmenting climate change issues across multiple institutions. This arrangement is typically favored by large, federal states that have traditionally not taken climate change issues seriously. Examples include Russia and the USA.

Coordinating climate change by establishing a division within an existing institution. Although their overall duties encompass a broader environmental remit, these institutions have specific divisions focused upon climate change that coordinate national climate change strategies. Examples tend to include developing countries that have fewer financial or expert resources, like Antigua and Barbuda, Belgium, Georgia, Kenya, Mexico, and Niger.

Merging broad, pre-existing environmental ministries with climate-related topics, like, for example, energy or waste. This type seemed to be the preferred response of parliamentary republics, for example, Austria, France, Germany, India, Italy, Trinidad and Tobago, and the UAE.

Each of these institutional arrangements has advantages and drawbacks. Having a specific institution devoted to climate change raises its political importance, but may make it harder to establish links with other areas of policy. Placing climate change within the remit of an existing institution lessens its power, but makes it potentially easier to influence broader policy. Merging pre-existing ministries with climate-related topics makes clear links between climate change and another area of policy, but may reduce its importance in the overall policy pecking order.

As Figure 3.1 shows, the proportion of countries adopting each of the different institutional strategies shows a clear preference for housing climate change in its own institution (32 percent), or merging it with an existing institution (23 percent). Fragmented approaches, whereby climate related policy remained spread across multiple departments, were relatively rare and existed only in the large federal republics of Russia and the USA.

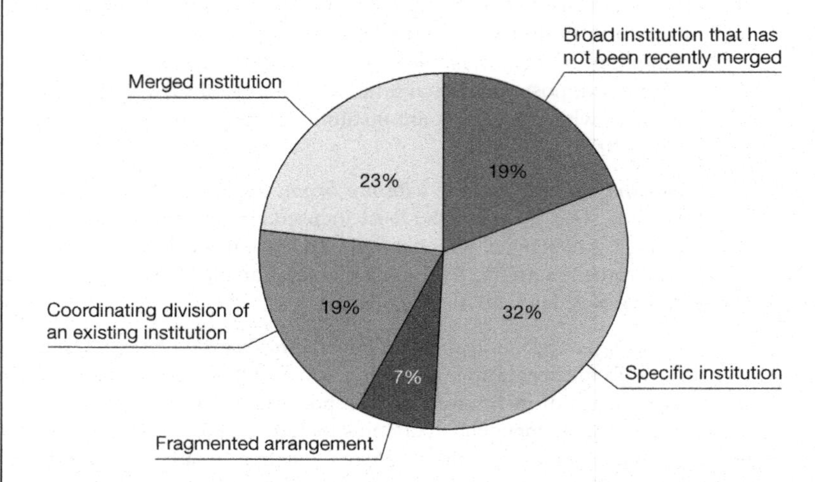

Figure 3.1 ***National strategies for institutionalizing climate change***

Overall, these trends indicate that climate change is being taken seriously by governments, but that there are wide differences in how this commitment is translated into an institutional framework. As Meadowcroft (2009) points out, this is not just dependent on the different political systems that characterize different countries, but also on factors such as the cultural, legal and administrative practices common to the country. The institutional arrangements adopted by different countries will exert a considerable influence over the importance that is attached to climate change as a policy issue, and shape how it is addressed in the future.

requirements of different actors and different contexts. Accordingly, rules tend to concern the procedures for making collective decisions, rather than determining the content of those decisions. This can include the role and position of different actors, the boundaries between them, who has overall authority, how interests are aggregated and the way in which information flows through the decision-making process.

Figure 3.2 ***Rules are critical***

Source: reproduced with permission from Thad Guy.

Ostrom *et al.* (1994) developed the institutional analysis and design framework to provide a structured way to think about how governance is conducted, by clarifying the different types of rules that are required in order to establish a working framework for cooperation. In their classic book *Rules, Games and Common Pool Resources*, Ostrom *et al.* (1994) distinguish seven types of rules that can be used to analyze institutions:

Position rules. These define what positions are available and how participants are assigned to positions. Position rules form the basic framework upon which other rules are superimposed, for example, the relative authority of each position.

Boundary rules. These specify the conditions under which participants may enter and exit various positions, and what their range of activity is when they are in them.

Authority rules. These prescribe how each position can act at various times, including the rights and responsibilities of each position, the resources available to them to support action, and their influence over other positions.

Aggregation rules. These set out how collective decisions are made, and the roles different position holders may play in reaching decisions.

Scope rules. These limit the range of possible decisions that may be reached, and ascribe status to decisions in terms of their importance and influence.

Information rules. These describe what information should be made available to each position at different times.

Payoff rules. These clarify how different actors either are, or are not, allowed to benefit or incur costs to themselves based on decisions taken.

The institutional analysis and design framework can be used to analyze and compare decision-making procedures across different policy areas, and provides a useful tool to understand the ways in which environmental institutions operate. As Figure 3.3 shows, the framework identifies four external factors:

The physical world. This includes the current characteristics and state of the resource in question, including human impacts upon it.

The attributes of the community within which the actor is embedded. This encompasses the various norms and common understandings concerning the resource in question.

The rules that enable and constrain action. This includes the sanctions for failure to follow rules.

The interactions with other individuals. This encompasses the micro-level interactions that may take place within an institution.

These factors impact upon the action arena, which comprises the participants in the decision-making process, who occupy different positions, and must determine a course of action based upon the information that they have, and their perceptions of the perceived costs and benefits. The way in which the participants interact will determine

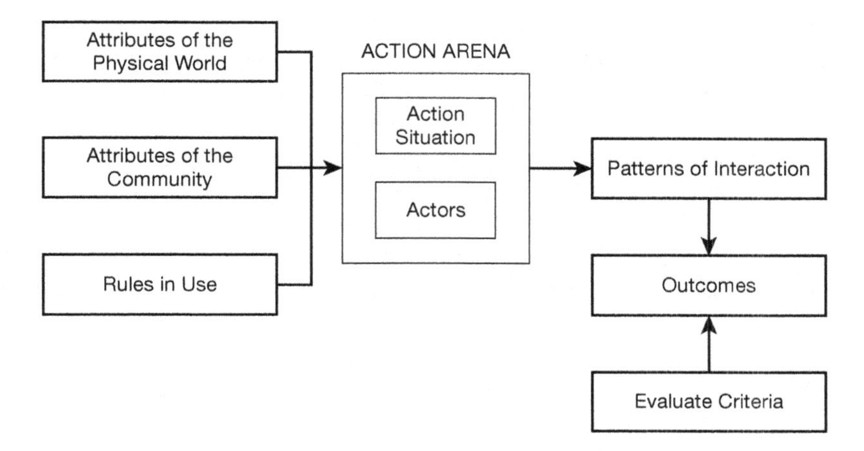

Figure 3.3 ***The action arena***
Source: adapted from Ostrom *et al.* 1994: 37.

the outcome of the decision-making process, although subsequent evaluation may alter it over time. The action arena provides the basic unit of analysis for the institutional analysis and design framework.

In addition to the seven types of rules identified above, Ostrom (1990) distinguishes between open rules, which can be debated and changed, and closed rules, which are set in stone. She also identifies a hierarchy of rules, ranging from operational rules that concern the way in which an institution operates, and collective choice rules that concern the ways in which decisions are made, to constitutional rules that frame the ways in which the rules themselves can be changed. Each of these levels has a different speed of function, scale, and generality. For example, operational rules can be changed fairly quickly in response to a new problem, whereas constitutional rules require a longer period of experience to justify changes to an institution's rule-making procedures.

The concept of "good governance" has emerged alongside governance as a set of principles to ensure that governance occurs in a democratic and fair way. It emphasizes the need for clear lines of authority in any decision-making process, to engender trust in the fairness of the process and accountability for resultant decisions and their consequences (Hyden 1992), and is largely about establishing transparent and accountable institutions to monitor rules and rule-making processes. Good governance has been used to promote specific causes, like establishing the conditions for free trade and reducing political corruption. The fate of the Advisory Group on Greenhouse Gases (AGGG), established by WMO, UNEP and ICSU in 1986 (two years prior to the IPCC), indicates the problems of not instituting a transparent decision-making process. The Advisory Group on Greenhouse Gases brought scientific experts into dialogue with policy-makers, but failed because its meetings mostly took place behind closed doors, and the motives of its largely philanthropic funding bodies were viewed with suspicion. This lack of transparency meant that the group lost scientific and political legitimacy and fell into disuse by 1990 (Agrawala 1999). By contrast, the IPCC, discussed later in this chapter, succeeded as an institution because it was scientifically and politically transparent with clear rules and procedures.

Common pool resource management

The tragedy of the commons rests on the assumption that individuals will act in their own self-interest to maximize the benefit they receive from using a common resource. Returning to Hardin's example of a piece of common grazing land used by multiple herders, the argument is that such behavior will destroy the resource over time, as each herder seeks to graze ever more cattle. Traditionally, it has been suggested that such selfish behavior must be corrected by the state, which can impose regulations to constrain the use of common resources. More recently, proponents of markets have suggested that the solution is to privatize common resources, to create an externally regulated market within which individuals bear the costs of their own over-use. Although presented as opposites, the state and market solutions are structurally similar in that they both assume that individuals act in isolation, and seek to avert the tragedy by imposing external controls (whether through laws or markets).

Ostrom offers an alternative to the external coordinating mechanisms of either markets or state regulation. Questioning the assumption that individual users act in isolation, her work shows how traditional societies that depend upon common resources like fisheries or grazing land develop internal coordinating mechanisms, which allow them to use common resources sustainably for long periods of time. For example, users might devise a collective set of rules for how a piece of grazing land can be used, which may include specifications for who can graze their herd, when, and to what extent.

Rules developed by the users themselves will often be more effective than those that can be devised by an external state organization.

As Ostrom (1990: 17) points out, "the herders, who use the same meadow year after year, have detailed and relatively accurate information about carrying capacity." For an external authority to develop such an in-depth understanding of a resource, including how it responds to different types and levels of use over time, would be, at best, hugely costly and time-consuming, and, at worst, impossible. How could an external agent hope to capture the experience of generations of herders?

Brian Wynne's (1996) classic study of radioactive fallout over upland Wales in the aftermath of the Chernobyl disaster discusses in detail how sheep farmers had a far greater understanding of the ecosystem than the scientists charged with deciding how long to ban their products for. While the scientists imposed a blanket ban on sales of lamb from the entire region, the actual level of contamination varied hugely depending on the underlying soil conditions, which in turn affected the grass that the sheep consumed. Failing to involve local farmers in the decision-making process led to a sub-optimal decision.

In addition to deeper levels of understanding, communities also buy-in implicitly to a system of self-imposed rules. An external agency (whether it be a market or a state regulator) would need to employ someone to enforce its rules, whereas, because any infraction will hurt them, the community of herders will ensure that the rules are followed and act together to monitor each other's use of the resource.

Ostrom suggests that the policy literature has tended to ignore the possibility that users can internally regulate resource use because such rules are often not obvious to the external observer. As Oran Young (1982: 18) notes, "social institutions may and often do receive formal expression (in contracts, statutes, constitutions or treaties), but this is not necessary for the emergence of or for the effective operations of a social institution." As noted above in relation to institutions, informal rules are just as important as formal rules, and inherited customs, traditions and deeper social structures, which may often appear unrelated to resource use, all play their part. As a result, "when the enforcement agency is not an external government official, some analysts presume that there is no enforcement" (Ostrom 1990: 18). Case study 3.2 gives an example of how common resource management works in practice, based on a community of fishers in Turkey that developed an internal set of rules to prevent over-fishing.

Case study 3.2

The Alanya fishery

Fikret Berkes (1986) describes how a community of about 100 fishers developed an internal set of rules to govern their use of a single inland fishery off the coast of Alanya in Turkey. Over-fishing in the 1970s had led to less predictable hauls and violent encounters between boats seeking to fish at the most abundant spots. Forming a cooperative group, the fishers devised a set of rules to protect the common resource and improve the predictability and size of each boat's annual catch. The system they devised allocated registered fishers to pre-identified fishing spots by drawing lots to decide which fishers got which spot. Starting in September, the fishers would then move one spot west every day until January, at which point they reverse and move one spot east until May. This movement tracked the migratory movements of the fish, ensuring that each fishing boat had a fair opportunity to fish each spot.

The system has a number of advantages:

- The fishing spots are placed far enough apart that the nets of one boat will not affect the catch of the neighboring boat.
- Resources are not wasted searching for a spot or fighting over the best spots, as they would be in a free market situation that simply allocated daily catch allowances.
- The fishers monitor each other, and any infractions are usually resolved at the local coffeehouse without the need for costly external intervention.
- The system of spots and movement is based upon decades of knowledge and experimentation by the fishers, maximizing the efficient use of the system in a way that external government regulation could not.
- The system reflects the dynamic nature of the resource, whereby fishing rights change daily to reflect the shifting distribution and quantity of fish. This kind of flexibility would be impossible under a system of private property rights, in which each fisher would own a particular spot.

As Ostrom (1990: 20) concludes, "Alanya provides an example of a self-governed common property arrangement in which the rules have been devised and modified by the participants themselves and are also monitored and enforced by them." Similar examples have been documented all over the world for community-managed resource systems like irrigation systems, communal forests, and hunting rights.

Classic responses to the tragedy of the commons are at fault because they assume that individuals are powerless to change a situation themselves. By contrast, Ostrom argues that analysts should focus on the internal and external factors that enable a community to self-organize and develop rules to sustainably manage common resources themselves. She suggests eight design principles to ensure cooperation:

Clearly defined boundaries. The resource system under management and the units to measure each individual's use of it must be clearly defined.

Proportional equivalence between benefits and costs. Each user should be allocated amounts of resource dependent upon the context of their needs.

Collective choice arrangements. All those affected by choices concerning a resource should be included in making decisions about concerning that resource.

Monitoring. Monitoring of the resource and resource users' behavior is essential, and the monitors should either be directly accountable to users or be users themselves.

Graduated sanctions. Penalties for misuse of a resource should be incrementally related to the degree of misuse.

Conflict resolution mechanisms. Where conflicts occur, resolution systems should be low cost and take place at the local level.

Minimal recognition of rights to organize. Because users have long-term rights and interests in the resource under management, they should be allowed to organize their interests as they see fit, which may include the formation of new institutions.

Nested enterprises. Where the resource under consideration requires the global coordination of local actions (for example, biodiversity conservation) the above design factors should be organized in nested layers at a range of scales.

Through these design principles, Ostrom applied her insights from traditional societies in the developing world to suggest how the use of common resources can be made more sustainable in the developed world context. Obviously this is not a straightforward task—global commons are far more complex than a single inland fishery, and a quick comparison of the atmosphere against a few of Ostrom's eight principles above immediately indicates the potential problems of scaling this model up. Neither the atmosphere itself nor its units of use are easy to define, as will be discussed in Chapter 6. In principle, collective choice arrangements would need to involve the entire world's population, as everyone is affected by the global warming that has resulted from misuse of the atmospheric commons—patently an impossible task. Getting users to monitor each other is tricky as greenhouse gas emissions are invisible, and the impacts of one user breaking the rules and emitting more than their fair share are not immediately detectable. The final point on Ostrom's list provides a partial answer to these problems—global commons must be managed as a set of nested enterprises, whereby local resource management systems are coordinated regionally, which are then coordinated globally—but there are a number of challenges to linking governance at different scales.

A note on scale

The way environmental issues play out at different scales is enshrined in the Rio mantra "Think global, act local," which reflects a commonly accepted hierarchy that runs from the local through the regional and national to the international levels (Kutting and Lipschutz 2009). This hierarchy underpins the idea of nested institutions, whereby each fits within a larger scale like Russian dolls. For example, the EU has a European Environment Agency that acts at the continental scale, while under this each country has its own national environment agency, which in turn will have regional branches. In this way, institutions can be nested to allow information to flow from the local to the international levels, and back again.

There is a huge literature on scale and it is worth considering briefly as it is an important concept in environmental governance (Sayre 2005). Social scientists have emphasized how framing problems at different scales can have major impacts on how they are subsequently dealt with. As Duffy (2006: 109) notes, it is important to examine the interplay between the national, the global and the local scales, as, rather than being "discrete and separable, they are inextricably interlinked." This linkage means that environmental problems can be framed at different scales to suit different purposes. For example, the global framing of many environmental issues, discussed earlier, leads to a "one-world" rhetoric, which privileges certain views and solutions over others. In his study of the construction of an amenity barrage across the Taff–Ely estuary in Cardiff, South Wales, Cowell (2003) shows how re-scaling the project to the supra-national level as a matter for the European Commission allowed local arguments over the loss of important ecological habitat to be reframed as a national rather than local issue, leading to a set of solutions based on habitat replacement elsewhere.

But this is not to say that actions at the local scale are necessarily better than actions at any other. Brown and Purcell have discussed (2005) what they term the "local trap" in environment and development, whereby it is simply assumed that devolving decisions and action to the local level is necessarily the fairest and most effective way to govern. Local governance arrangements are often constrained by pre-existing tensions and power dynamics between key stakeholders, and fail to benefit from the economies of scale that accrue when problems are addressed at higher levels. There is a thin line between ensuring that governance unfolds in a way that is appropriate and sensitive to local contexts, and reinventing the wheel, often ineffectively as resources have been spread too thinly.

One of the biggest challenges facing environmental governance involves overcoming the mismatch between political and ecological scales. So, for example, watersheds comprise coherent ecological units within which resource management issues like fishing, water extraction and pollution are best dealt with. Historically speaking though, rivers have more often been used as political boundaries, splitting watershed management across different jurisdictions. Many environmental governance initiatives seek to establish jurisdictions that are organized on the basis of coherent ecological units, like the EU Water Framework Directive which brings stakeholders together to co-manage watersheds (White and Howe 2003). At a larger scale, concepts like bioregions and even city-regions seek to align political units more closely with the ecological and economic spaces on which they depend. But to date, attempts to govern at ecological scales tend to overlay rather than replace pre-existing political scales. Power and legitimacy remain invested in the traditional political institutions, making it hard to overcome established scales of governing (Sneddon 2002).

Some theoretical frameworks reject the importance of scale altogether. For example, actor network theory, discussed in Analytics of governance 3.1, suggests that there is no such thing as scale, merely networks of humans and non-humans that are linked together in different ways. Bulkeley (2005) explores the theoretical complexities of this debate, discussing how scale and network concepts have been used in studies of environmental governance.

Analytics of governance 3.1

Actor network theory

Actor network theory (ANT) sees the world as being constituted of networks, rather than as networks being an abstract description of the world. In *We Have Never Been Modern*, Bruno Latour (1993) narrates Louis Pasteur's discovery of yeast in 1857 as the story of how a new network was formed between the bacteria, the microscope, the scientist, and, subsequently, the aristocratic benefactors of his research and the learned societies in Paris to whom he demonstrated his discovery. The key insight of ANT is that the bacteria were altered in the act of being discovered by the Frenchman's gaze, as the very act of observing them down the microscope enrolled them into a new network of people and things.

ANT extends the ability to act to non-humans as well as humans. Hinchliffe's (2001) account of the British BSE epidemic ("mad cow disease") places the prions, whose exact status within cow brains scientists were unable to determine, at the center of the crisis. Examples of non-human actors abound in the environmental sphere, from charismatic species like polar bears (who exert considerable influence over public responses to climate change), to food (which rots), water (which flows, sometimes floods and is incompressible) and, most recently, carbon (whose movements in and out of various ecosystems remains uncertain). Focusing on the ambivalent and often unknowable behavior of non-humans allows uncertainty to be given of ontological status as an actor in its own right, rather than simply being cast as an epistemological irritation that can be wished away with error bars, or that will be defeated by the indomitable march of progress. It thus provides a rich analysis of environmental issues that involves non-humans as subjects rather than just objects of governance.

Returning to the topic of scale, ANT is based upon a relational ontology, which means that it focuses on the connections between actors rather than their actual distribution in space. In other words, organizations may be located on opposite sides of the planet, but if they are connected through various personal acquaintances, bonds of trust and regular cooperation together, they are in effect much closer than organizations that occupy the same building, but who are unaware of one another. Relational ontology dispenses with the concept of scale altogether, as it simply has no meaning when one talks solely about connections. The classic example given is that of a railroad, which can be seen as local, regional or national, depending on the scale at which it is looked at. In this sense, the railroad itself has no scale. Rather than requiring complex concepts to understand how different scales are linked, ANT simply focuses on who or what is connected.

Key actors

If rules and institutions set the stage for governance, then it is now possible to introduce the main actors. This section considers the state, society, business, supra-national organizations, NGOs, international scientific advisory bodies, and sub-national actors. The discussion of each actor is not exhaustive, but is intended to provide necessary background for the subsequent chapters.

The state

Whether the state is seen as a critical player or not within different modes of governance, some understanding of what it is and how it functions is necessary to understanding any form of governing. Many political and social theories see the state as simply reflecting the interests of particular parts of society. For example, theories of pluralism emphasize the way that numerous different social interests are represented and furthered by the state, while neo-Marxists see the state as furthering the interests of the bourgeoisie (middle classes). Structuralists define the state as a system for aggregating interests and producing policy, but are more interested in the political processes by which this occurs than the state itself (Kjaer 2008).

In contrast, state-centered theories see the state not as being reducible to specific social interests or a political system, but as constituting an autonomous actor in its own right (ibid.). Neo-statists argue that the autonomous capacity of the state is a product of its differentiation and specialization, which gives it the capacity to formulate and implement policy across a broad range of fields. Echoing the insights of new institutionalism, neo-statists argue that policy often originates within the state itself, as networks of officials and bureaucrats adopt new thinking over time, rather than simply in response to the changing demands or needs of external social or economic interests (Almond 1988). Although it is common to speak about "the state" or "the government" as if they are single, monolithic entities, their structure and activities are in reality highly fragmented, comprised of multiple departments and agencies with often overlapping powers and responsibilities (Jones and Evans 2006). The discussion in this book does not go into detail concerning how various modes of governance might be operationalized by states, as this would venture into the domains of policy studies and public management, but it does contrast the different roles for the state that each mode of governance implies.

Because governance involves the state and non-state actors in governing, a long-running debate has developed in the literature concerning the exact role of the state. According to advocates of network governance, the role of the state is not only reduced, but radically reconfigured as it becomes just one of a number of stakeholders involved in governance. But other authors question whether state-led governance has really been superseded (Davies 2002). While appealing in the abstract, examples spring to mind that suggest the crisis of state governance may be overstated. National governments still exert primary control over their populations, whether through setting education, welfare or health agendas, or regulating immigration. The Weberian state bureaucracy is not dead, and still embodies many normative features of democracy and good administration. Further, state policy shapes commercial activity; as Hawkins (1984) notes, "the power to define and enforce consents is ultimately a power to put people out of business, to deter the introduction of new business or to drive away a going concern." Perhaps most poignantly, as the financial crisis of 2008 demonstrated, even the private sector is dependent on the resources of the state in the final analysis. Governance networks contain the inherent ability to fail, and when they do it is the state that picks up the pieces (Jessop 1999). The institutions of government have changed enormously, as have the means by which they achieve their ends, but in many cases states still control the rules of the game (Pierre 2000, Pierre and Peters 2000).

Civil society

Civil society plays a key role in environmental governance. Non-point source pollution like tailpipe emissions is essentially produced by society at large, and thus to address it, it is necessary to engage society in governing (Landy and Rubin 2001). Sustainable development emphasizes the normative idea that citizens should have the ability to influence how the places in which they live are managed, emphasizing local action and community inclusion. The public also have valuable knowledge about their own environment, which, in specific localities, can be more accurate than that of scientists or external experts. As Irwin (1995) has argued, there will be no sustainability without a greater potential for citizens to take control.

The benefits of community self-governance that Ostrom identifies in relation to common resources also apply when communities engage with broader environmental issues that may be affecting them, both locally and beyond. Emanating from North America, civic environmentalism is one of the more well-developed schools of thought that exhorts local communities to "think locally and act locally" to address the environmental issues that are important to them. In one of its most well-known treatises, John (1994: 7), says, "the central idea animating civic environmentalism is that in some cases, communities and states will organize on their own to protect the environment, without being forced to do so." Similarly, Landy and Rubin (2001: 7) state, "in the real world, people don't view themselves simply as property owners or as consumers, but as neighbors, friends, parishioners, and citizens." The "civic" component of civic environmentalism suggests that people will get involved not because of some "founding environmental ethic or a commitment to the state but rather a responsibility stemming from their embeddedness in place" (Karvonen and Jocum 2011). Engaging people also raises their awareness of the value of the environment. As recent research shows, engaging in local, environmentally friendly activities like attending farmers markets generates a greater sense of ecological citizenship among people (Seyfang 2006).

Business

The goals of business and the environment are often presented in zero-sum terms, whereby economic growth automatically harms the environment (Welford and Starkey 1996). Undoubtedly there are many examples where the extraction of raw materials has devastated landscapes. Similarly, industrial production processes often have excessive energy and water demands, generating pollution in the form of chemicals, air emissions, waste water, solid waste, noise, dust and odor, and, in terms of their final disposal, products themselves. Socio-economically, a business impacts on the environment in terms of jobs and the well-being of its workforce. One does not have to look far to find examples of environmental disasters caused by private companies. Taking two contrasting examples, the 2010 BP deep sea oil leak into the Gulf of Mexico could cost the company in excess of $30 billion, while the Minamata Bay disaster in Japan, which saw chemical industries dump mercury into the sea between 1930 and 1960, led to thousands of people developing methyl-mercury poisoning through the consumption of contaminated fish.

Given their massive impacts on the environment, businesses are key players in environmental governance, which is one reason why sustainable development is based upon the premise that environmental protection and economic development can occur simultaneously to produce a win–win scenario. Good environmentalism is good business, it is suggested, because companies can save on production costs and enhance their public image (Welford and Starkey 1996). In countries such as Germany and the Netherlands, big business has worked closely with government to steer environmental policy under a model of governance called ecological modernization, discussed in Analytics of governance 3.2.

The greening of business and industry has passed through a series of stages. The 1960s and 1970s were characterized by outright denial that any environmental problem existed, which was addressed by tightening the centralized regulation of business in the 1980s. The 1990s saw businesses become increasingly compliant towards environmental regulation, with environmental leaders emerging in the 2000s who sought to go beyond the minimum legal requirements (Berry and Rondinelli 1998). Businesses are responding to market pressure from customers and investors, regulatory pressure from governments, and social pressure from NGOs and the public to improve the social and environmental impacts of their operations. As a coercive form of governing, governance builds upon this latest stage, seeking to encourage businesses to voluntarily engage in more environmentally friendly operations.

Supra-national organizations

Supra-national organizations, which bring nation states together, play a crucial part in coordinating collective action at the global level. Among these, the UN has played the central role organizing international environmental conferences and hosting many of the organizations and secretariats that administer international environmental agreements.

Established in 1945 to replace the League of Nations, the stated goal of the UN is to maintain peace and security through international cooperation, help solve international economic, social, cultural and humanitarian problems and promote respect for rights and fundamental freedoms. What British sociologist Anthony Giddens (1990: 139) famously called "the runaway juggernaut" of globalization was actively pursued by the UN in the aftermath of World War II in order to make countries more interdependent and thus prevent another world war

Analytics of governance 3.2

Ecological modernization

Emerging in the 1990s, ecological modernization recognized that the future health of industry depended on maintaining a "sustenance base" of environmental resources (Mol 1995). Rather than strong government regulation, it suggested that environmental improvements could be achieved by big business working closely with governments to shape environmental policy (Fischer and Freudenberg 2001).

Ecological modernization is technocentric, relying heavily on scientific research and technical expertise to develop new technologies that will enable more environmentally friendly economic growth. Its approach is managerial, comprising voluntary procedures and forms of self-regulation rather than legalistic and adversarial state-led regulation. The pragmatic emphasis upon solving problems with industry gives officials considerable scope to interpret policy, and demands new, more accommodating political structures (Spaargaren 1997). Policy development and implementation is corporatist, occurring in close consultation with the industries that it is intended to influence. As Dryzek (1997: 144) states, ecological modernization "implies a partnership in which governments, businesses, moderate environmentalists and scientists co-operate." Regulations are often defined in terms of economic feasibility as well as technical feasibility, bringing environmental policy into closer union with economic policy to encourage technological innovation (Young 2000).

By its very nature, governance implies a pluralist state with lots of interests involved in policy-making, but critics suggest that ecological modernization creates a neo-corporatist state where business interests are overly privileged. Indeed, some have attributed the success of ecological modernization to its amenability with right-wing policies that favor the interests of private industry, while others have suggested that the theory actually just reflects developments in Germany and the Netherlands in the 1980s and 1990s and has little relevance elsewhere. That said, most governments now develop industrial regulation in close consultation with industrial leaders and consult closely before taking decisions that may impact business.

(and whatever else globalization might stand accused of, it has succeeded on this front).

The UN has six primary organs: the General Assembly (main meeting assembly), the Security Council (concerned with maintenance of peace), the Economic and Social Council (to coordinate the work of the UN in these fields), the Secretariat (providing information and support to the rest of the UN), the International Court of Justice (the judicial arm) and the Trusteeship Council (which is now redundant). These six are complemented by a number of specialized agencies, such as UNESCO, FAO, and WHO, which were set up to coordinate work and establish international links in specific fields, like education, food and health, where there were obvious needs (Speth and Haas 2006).

Initially the work of the UN was mostly normative, concerned with setting agendas and facilitating cooperation, rather than engaging on the ground, but this changed as the number of specialized agencies and subsidiary bodies grew in response to the needs of global cooperation and development as perceived by the General Assembly. There are now 17 specialized agencies, which report through the Economic and Social Council to the General Assembly, and more than 12 subsidiary bodies which report directly to the General Assembly as well as to the Economic and Social Council. Within these groups there is a tremendous diversity in terms of size and programs. The Economic and Social Council has also created regional subsidiary bodies (for example, the Economic Commission for Africa, or the Economic Commission for Europe) as well as standing committees on topics such as natural resources and science and technology for development.

As the most important global environmental institution, the origins of the United Nations Environment Programme (UNEP) are important in understanding how current environmental governance operates. Established in the aftermath of the Stockholm Conference on the Human Environment in 1972, there was little agreement on either what a global environmental institution should do, or how it should be set up. Scandinavian countries favored the formation of an environmental council within the United Nations, while others (including the UK, the USA and France) opposed the creation of a strong, independently funded agency, preferring a program. UNEP's primary role was catalytic, acting as what the third general secretary of the United Nations, U Thant, called a "switchboard" organization that would coordinate and facilitate the environmental work of other UN agencies. But he also stated the need for UNEP to be strong enough to "police and enforce" its decisions. Officially, its role is to support coherent international decision-making processes for environmental governance by:

- Providing an international framework for environmental politics.
- Developing international environmental databases.
- Establishing a series of environmental agreements.

While UNEP is relatively small in comparison to many UN bodies, it has provided the spark for a series of successful international environmental agreements, from the Convention on International Trade in Endangered Species (CITES) in 1973, which protects the trade of endangered species, to the Agreement on Long-Range Transboundary Air Pollution in 1979.

Despite notable successes, criticisms of UNEP abound. Some argue that its brief is split confusingly between political and scientific goals, while most agree that it lacks the resources and staff to coordinate effectively. One could argue that the first problem is endemic to any organization working in the environmental field, as it overlaps with so many other areas of concern. But the second is more specific and relates to the problem of being a program rather than an agency. This means that it is funded by donations rather than an allocated budget. Not only is precious time spent courting donations, but promised donations can fail to materialize. Although its annual budget has grown from some $20 million in its first year to $120 million in 2003, less than 4 percent of this total comes from the United Nations. The result is that UNEP can only focus on a certain range of issues and initiatives for a limited time, leading to activities that can appear haphazard.

It is interesting to compare the United Nations, which has 192 member states out of the 245 countries currently recognized, to the EU, which has 27 member states and constitutes the most important regional supranational organization in the environmental sphere. The origins of the EU are also found in the need for stability after World War II, with the Treaty of Rome establishing the European Economic Community in 1957, and the Maastricht Treaty creating the EU in 1993. Unlike the United Nations, the EU has a legal mandate from its member states to protect the environment and deliver sustainable development, coordinating environmental policy across it member states in order to ensure that there is a fair playing field for economic competition (Axelrod *et al.* 2005). It is the largest producer of environmental policy in the world and has piloted innovative governance schemes, like the Water Framework Directive that requires governments to draw up River Basin Management Plans for their major watersheds. The EU issues framework directives, which set out common goals but leave members room to decide how to meet them (Jordan 2002). Unlike UN agreements, member states of the EU cannot choose whether to opt in or out, but are legally bound to implement framework directives or face hefty fines. Additionally, the EU is funded directly by member states, and has democratically elected members of parliament.

Although the EU cannot coordinate global action like the UN, it votes as a block in environmental treaty negotiations, making it a powerful diplomatic broker. While Europe may not have the military or economic influence of the USA or China, it has a potentially important role to play in acting as a global exemplar for environmental reform (German Advisory Council on Global Change 2009).

NGOs

Because NGOs are in many ways a product of governance it is unsurprising that they have a key role to play in facilitating collective action, and many of the case studies discussed in this book involve NGOs. While it is hard to imagine the political landscape today without NGOs, they only came into being in the immediate post-war period, engaged primarily by international institutions like the UN to help implement programs and respond to humanitarian emergencies. The UN coined the term NGO to differentiate between the public inter-governmental bodies and private international bodies with whom they worked (Willetts 2002). Since the end of World War II, the number and diversity of NGOs has exploded, and, as representatives of civil society, they are integral to the philosophy of modern governance, which prioritizes the inclusion of non-state actors in order to enhance the legitimacy of decisions. Gemmill and Bamidele-Izu (2002) are probably not guilty of hyperbole when they state that, "the very legitimacy of international decision-making may depend on NGOs as a way to ensure connectedness to the publics around the world and substitute for true popular sovereignty, which international bodies, devoid of elected officials, lack."

NGOs are massively influential in the environmental field, and many enjoy very high profiles. Greenpeace and Friends of the Earth are both global NGO charities with household names, but they began life as protest groups. In the case of Greenpeace, this involved a group of anti-war protestors from west coast America chartering a battered fishing vessel and sailing through the US nuclear weapons testing area on Amchitka Island off the coast of Alaska in 1971. Their actions captured the public imagination, and set the tone for subsequent environmental NGOs who played a key role in bringing such issues to the attention of politicians. The transformation of environmentalism from counter-culture to formal policy concern can be told as a story of successive NGO campaigns, on issues from desertification to climate change, that were often explicitly designed to provoke popular and political action.

Today the exact number of NGOs is unknown, but it is substantial. The Philippines alone has 18,000 NGOs, while the European Environmental Bureau, which acts as a match-maker between environmental NGOs and various parts of the EU, has 132 European members, representing 14,000 member organizations and 260 associate organizations. The staggering growth of NGOs is related to the rise of governance itself, the sheer complexity of environmental issues, and developments in communication technologies that have facilitated cheap and effective networking (McCormick 2005).

NGOs perform five major roles in environmental governance (Gemmill and Bamidele-Izu 2002):

- Collecting, disseminating and analyzing information.
- Providing input to agenda-setting and policy development processes.
- Performing operational functions.
- Assessing environmental conditions and monitoring compliance with environmental agreements.
- Advocating environmental justice.

NGOs have varying levels of involvement in governance and play a number of roles, consulting to government or industry, drafting treaties, and even regulating activities (Charnowitz 1997, Cashore 2002). They act on behalf of their members (although how democratic they are varies) but also as significant political pressure groups in their own right, often contributing directly to national and international policy-making (Betsill and Corell 2008). One of the reasons NGOs are valuable partners is that they can do things that governments and private companies simply cannot. "By supplementing, replacing, bypassing, and, sometimes, even substituting for traditional politics, NGOs are increasingly picking up where governmental action stops—or has yet to begin" (Princen *et al.* 1994: 228), stepping in where there is a gap that governments or companies are unable fill.

NGOs can respond rapidly in specific localities because they have pre-existing contacts on the ground. They can also help in countries where for one government to directly aid another may be viewed as politically undesirable by one or both sides (Simmons 1998). Similarly, NGO-run environmental networks may prompt companies to join to be seen to be keeping up with their competitors, where they would not be able to justify individual action to their board or shareholders. They also provide an acceptable substitute for direct state regulation, for example, monitoring private compliance with environmental agreements.

But while governments and international institutions seek to involve NGOs in governance, the mechanisms by which NGOs engage and are engaged are largely informal and unregulated, creating a danger that NGOs over-represent special interest groups. For example, the emphasis on tropical forests in global environmental governance is largely a result of lobbying in the 1980s by developed world NGOs, whose membership was obsessed with rainforest protection (Humphrey 1996). In terms of climate change, peat-lands actually represent a larger global carbon sink, but, lacking strong promotion from NGOs, have barely made it onto the global agenda (Joosten and Couwenberg 2008).

Furthermore, it would be wrong to view the global community of environmental NGOs as sharing common goals and methods—for example, the political tensions between the developed and developing worlds are reproduced in the global NGO community (McCormick 2005). NGO funding can also be opaque, but perhaps no more so than academic organizations like universities, which rely on the government in addition to many other donors.

There are dangers to constraining NGO involvement in governance, as their very strength lies in their diversity and creative ways of networking, but, when involved in major decision-making, their selection and workings need to be transparent and open to scrutiny. In an attempt to mitigate this problem, the Commission of Sustainable Development has identified eight major groups within civil society (women, children and youth, indigenous peoples and communities, non-governmental organizations, workers and trade unions, the scientific and technological community, business and industry, and farmers), which are intended to ensure representative diversity when they engage NGOs in governance.

International scientific advisory bodies

International scientific advisory bodies represent a mechanism through which the advice of scientists on environmental issues has been institutionalized and formalized for decision-makers (Biermann and Pattberg 2008). They provide overviews of different elements of the planetary system, such as the atmosphere and biosphere, presenting cutting-edge scientific knowledge about the global environment.

For example, the Millennium Ecosystem Assessment resulted from a meeting at the World Resources Institute in 1998, which identified major gaps in knowledge and understanding of global environmental resources. Given the potentially disastrous consequences of ecosystem change for human well-being, and the fragmented nature of global conservation data, it was suggested that a new international assessment process was required to produce an overview of the Earth's ecosystems. Launched in 2001, the assessment brought together more than 1,360 experts from NGOs, academic and research organizations across 95 countries to produce five technical volumes and six synthesis reports on the world's key ecosystems. The assessment provided a broad consensus view of scientists that was intended to form a basis for decision-making, but also identified areas with insufficient information to reach consensus.

As well as providing a global scientific overview, international scientific advisory bodies work closely with the international policy community, which can create tensions between scientific and political modes of establishing truth. Case study 3.3 discusses this issue in relation to perhaps the most influential international scientific advisory body, the IPCC.

Sub-national actors

Recent research has highlighted a growing diversity of actors and sites involved in governance below the national level, as regions, localities and cities begin to deploy their own strategies to engage with environmental governance (Bulkeley and Moser 2007). Betsill and Bulkeley's (2004) work on cities and climate change demonstrates the growing influence of cities as autonomous political units in tackling environmental issues. Taking the Cities for Climate Protection network as their case study, they show how cities are forming transnational alliances to address the emission of greenhouse gases and find solutions to urban sustainability, by-passing the national level of governance in the process (although, as Betsill and Bulkeley point out, their effectiveness remains constrained by the lack of resources that characterizes local government more widely).

In her study of Seattle in the northwest USA, Jennifer Rice (2010) argues that the city has asserted its authority as a coherent space in which to tackle climate change in three ways. First, it has "climatized" the urban environment, making climate change the driving force behind the city's whole approach to planning. Second, it has "carbonized" urban governance, developing greenhouse gas inventories and targets for all

Case study 3.3

The Intergovernmental Panel on Climate Change

More than 2,000 scientists from 154 countries typically participate in the IPCC process, which operates on a cycle of reports that are compiled and then released to governments at key moments before major international negotiations. Accordingly, the 1990 report was prepared in time for the Rio Earth Summit in 1992, the 1995 report in time for Kyoto in 1997, the 2001 report in time for the Johannesburg Earth Summit in 2002, and the 2007 report in time for the Copenhagen Climate Change Conference in 2009.

The IPCC was formed in 1988 to replace the Advisory Group on Greenhouse Gases, itself only brought into existence in 1986. The Advisory Group on Greenhouse Gases was felt to be too removed from the policy-making process, constituting an elite committee of scientists that was funded by politically motivated philanthropic foundations. The IPCC, by contrast, was composed mainly of people who participated not only as scientific experts, but as official representatives of their governments. It was thus conceived as an explicitly hybrid institution capable of producing a political consensus around scientific knowledge concerning the global climate (Weart 2008).

In aiming to produce reports that are policy relevant but not policy prescriptive, the IPCC makes great efforts to emphasize scientific integrity, objectivity, openness and transparency in its working methods. Reports go through a rigorous review process that involves experts around the world and all member governments. In terms of boundary rules (who can take part), membership of the IPCC is open to all member countries of the United Nations Environmental Programme (UNEP) and World Meteorological Organization (WMO). Scientists are independently nominated to be panel members by their own governments, and, in order to ensure the support of the developing world, each chapter of the assessment reports has a lead author from both the developed and developing worlds. This quota system to ensure North–South parity is more akin to a political than a scientific body, but is seen as a necessary prerequisite to secure collective action on climate change (Biermann and Pattberg 2008).

In terms of position and aggregation rules, 122 coordinating lead authors and lead authors, 515 contributing authors, 21 review editors, and 337 expert reviewers contributed to the Third Assessment Report. The reports are reviewed by governments as well as experts, and must be unanimously agreed upon by every member state and by all leading scientists serving as lead authors. These rules are designed specifically to produce consensus and buy-in; by getting member states to sign off each report, it is hoped that they will support the recommendations made at each international meeting. While this process has been criticized by many for watering down the IPCC's scientific recommendations, it has provided a politically credible, robust basis for collective action in the face of considerable uncertainty and major vested political interests. Despite muffling scientific experts to some degree, the IPCC has become more strident with its recommendations over time. The 2007 synthesis report, which combined findings from the four separate reports, made a clear statement that there was a 90 percent likelihood of disaster in a business-as-usual scenario.

Beyond the legitimacy generated by involving governments in the IPCC process, there are very practical reasons for doing so too. As Oberthür and Ott (1999: 300) note, "virtually no one involved in the negotiations is capable of grasping the overall picture of the climate negotiation process." The so-called "complexity trap" of scientific and legal technicalities necessitates a continuous dialogue between scientists and politicians to ensure that all interests and information are brought to bear upon the issue.

government activities that will make public activities carbon neutral. Finally, it has "territorialized" carbon, creating discrete geographical areas within which emissions are monitored and targets set. Territorialization plays an important role in empowering actors, as they can directly influence emissions in, for example, a neighborhood and monitor the effects of their action. Of course, Seattle has a few built-in advantages over many other places; 90 percent of its power supply comes from hydro-electric sources that are almost emission free, and it has a long legacy of environmental concern, which means public and institutional support is forthcoming. But the three-fold strategy used in Seattle applies in principle to actors at all levels who are seeking to establish effective climate change governance.

Regional initiatives have also sprung up to address climate change (Benson 2010). The Western Climate Initiative, established in 2007, includes the US states of Arizona, California, Montana, New Mexico, Oregon, Utah, Washington, and the Canadian provinces of British Columbia, Manitoba, Ontario, and Quebec. The initiative aims to lower emissions by 15 percent of 2005 levels by 2020 through a mandatory cap-and-trade system that is scheduled to begin in 2012. Power generation, transport and industrial emissions are all planned to be included in the market. The scheme is far from a gimmick, with members of the initiative responsible for 20 percent of total US emissions, and 73 percent of total Canadian emissions. Similar plans exist in the US Midwest (the Midwestern Regional Greenhouse Gas Reduction Accord agreed in 2007) and East Coast (the Regional Greenhouse Gas Initiative agreed in 2005).

It is no coincidence that many of the best examples of sub-national climate governance come from the USA, given its conspicuous lack of leadership on climate change at the national level, but urban and regional action on climate change can be found all over the world.

Conclusions

This chapter has discussed the role of institutions and rules in enabling collective action, before considering the key actors in environmental governance. Theories of institutionalism help understand the role institutions play in framing the possible range of actions available to their members. Within institutions, formal and informal rules steer collective action, and it is possible to identify the principal characteristics of rules that tend to enable the sustainable governance of environmental resources. Scaling these up from communities, in which rules often exist as traditions and customs, to address large-scale environmental problems effectively and legitimately represents a key challenge. Further, the requirements of rules and institutional design will vary according to the demands of different modes of governance.

The chapter then moved on to consider the key actors involved in environmental governance. Under governance, the institutions of government have changed enormously, as have the means by which they achieve their ends, but states still exert considerable power in setting the policy framework within which environmental governance takes place. Society was identified as a critical actor in environmental governance, as both the source of non-point source pollution and action at the local level, as were businesses, given their massive impacts on the environment. A key part of governance involves engaging the state, society and business voluntarily in the process of governing.

In addition to these three groups, a number of institutions have emerged as part of the broader shift to governance. At the international level, supra-national institutions like the UN are important in bringing nation states together to address environmental issues. At the sub-national level, actors like cities and regions are reterritorializing governance, highlighting the need for linkages to coordinate action between scales. NGOs are in many ways a product of governance, stepping in to perform duties that states cannot, and play a key role providing the glue between elements of civil society and supra-national organizations. In a similar way, international scientific advisory bodies, which represent a mechanism to achieve political consensus around scientific knowledge, provide institutional glue between the global scientific and policy communities.

Having set the framework and considered the key actors, the next chapter turns to the question of global environmental governance.

Questions

- Who creates and designs institutions?
- Design a set of rules to govern an institution that would represent your class.

Key readings

- McCormick, J. (2005) "The role of environmental NGOs in international regimes," in N. Vig and R. Axelrod (eds) *The Global Environment: Institutions, Law and Policy*, London: Earthscan, 52–71.
- Ostrom, E. (1990) *Governing the Commons: The Evolution of Institutions for Collective Action*, Cambridge: Cambridge University Press.

Links

- www.youtube.com/watch?v=ByXM47Ri1Kc. Elinor Ostrom talks about sustainability and the tragedy of the commons on this video produced by the Stockholm Environment Institute.
- www.ipcc.ch/. Home of the Intergovernmental Panel on Climate Change and its state-of-the-art reports on climate change.

4 Global governance

Intended learning outcomes

At the end of this chapter you will be able to:

- **Understand the process and architecture through which global environmental governance unfolds.**
- **Identify the key conferences, institutions and agreements relating to the environment and assess their legacies.**
- **Appreciate the importance of implementation and the challenges that it presents.**
- **Discuss the key debates surrounding the institutions of global environmental governance.**

Introduction

While the environmental issues facing society today are thoroughly global in character, there is no single institution, set of rules or overarching social contract or framework of cultural values through which to coordinate a response. The current system of global governance has evolved in the absence of a coherent political body, creating a situation in which institutions are charged with making global policy that countries will agree to follow without a global polity, or political body, to enforce them (Hajer 2003).

This chapter focuses on the three core elements of global environmental governance:

Process: the international meetings that are organized to address environmental issues.

Architecture: the institutions that are created to enact the agreements reached at these meetings.

Implementation: making sure that what gets agreed is put into practice.

Particular attention is paid to the meetings organized by the United Nations, from the Stockholm Conference on the Human Environment in 1972, to the Copenhagen Climate Change Conference in 2009. The chapter assesses the successes and failures of each of the three elements, and considers the key debates surrounding the institutions of global environmental governance, before discussing the key challenges facing them.

Process

While states can act unilaterally, i.e. on their own, the transboundary nature of environmental issues means that multilateral, i.e. collective, action is almost always required to address them. Global environmental governance is driven primarily by global meetings, which are organized to coordinate multilateral responses to environmental issues. Meetings are generated in response to the concerns of the international community, which are driven in turn by public opinion and/or the international scientific community. More than anything else, the international meetings organized by the UN have established the environment as a formal political concern for governments around the world.

Rather than each state having to negotiate with every other state, institutions like the UN can set out a commonly agreed set of rules for negotiation, making the process of international cooperation more efficient. Because all states agree to abide by the same rules, international political outcomes reflect a consensus, rather than simply the will of the most powerful state. The neo-realist model of international relations, discussed in Analytics of governance 4.1, sees the role of institutions as providing focal points for nation states.

Considerable information gathering is usually conducted by an international negotiating committee prior to potential meetings. Sometimes this process leads nowhere, while in other cases an issue can assume considerable importance, prompting a series of meetings over a number of years or decades. Without any overall guiding strategy, the process by which environmental issues get addressed can appear rather haphazard, and, as Case study 4.1 discusses, international bodies must sometimes take advantage of political conditions in a fairly opportunistic manner to secure agreements.

Analytics of governance 4.1

International relations

International relations studies how nation states interact, including the roles of intergovernmental organizations, NGOs and multinational corporations. A number of competing theories exist within international relations that explain the behavior of states in different ways, among which the most influential are the realist, neo-realist and liberalist schools of thought. Each has a bearing on global environmental governance.

The realist theory of international relations focuses on the system of nation states established originally in the Treaty of Westphalia in 1648, which is structured around the principles and instruments of international law. Nation states are like billiard balls—discrete legal entities that act in their own separate interests, whose relations with one another are defined primarily by the military force with which they can protect their interests. There is little or no real cooperation between states, simply balances of power. In focusing exclusively on states, realists are not concerned with governance *per se*.

Neo-realist models of international relations have developed the concept of regimes to explain the fact that nations actually do cooperate, albeit in a fairly disorganized way. The regime concept describes the broad social institutions, conventions and understandings that arise between nations relating to a particular issue, like nuclear proliferation, and the set of treaties that pertain to it (Speth and Haas 2006). Regimes are the "implicit or explicit principles, norms, rules and decision-making procedures" (Krasner 1983: 2) that allow nations to cooperate. Neo-realists seek to explain an international state of affairs that is more anarchic than that which would result from states acting purely in self-interest. This model addresses issues of governance because the concept of regimes extends realist theory to include actors other than states, viewing institutions as focal points for cooperation between nation states.

Liberalist scholars take this model a few steps further, arguing that non-state actors like NGOs are actually the most critical players in international relations. As McCormick (2005) notes, this school of thought is essentially idealist, in contrast to the realists, because it suggests that international relations are governed by ideas, or common interests, rather than self-interest. For liberalist scholars, the word "global" has a different meaning to "international" or "intergovernmental," going beyond states to encompass global institutions and civil society (Falk 1995, Rosenau 1995). The liberalist theory of international relations applies the principles of governance to the global stage, recognizing the increasing participation of actors other than states in rule-making and implementation ("multiactor governance"), and the emergence of new forms of organization such as public–private and private–private partnerships. In contrast to the billiard balls of realism, the metaphor of the cobweb is often used to emphasize the interdependence of states in the liberalist model.

Each theory of international relations explains part of the way in which environmental governance unfolds on the global stage. The question of whether global environmental governance really is governance beyond the state, or simply a system of intergovernmental negotiation, is returned to in the conclusion.

Case study 4.1

The Aarhus Convention

The Aarhus Convention on Access to Information, Public Participation in Decision-making and Access to Justice in Environmental Matters was negotiated in 1998, coming into force in 2001. Not only is it the only coherent treaty on public participation in environmental decision-making, but it was negotiated by the United Nations Economic Commission for Europe, which works primarily in the non-member countries of the EU in Eastern Europe and Central Asia, rather than by an environmental institution in Western Europe or UNEP.

The explanation for these oddities lies in the specific political conditions found in Eastern Europe in the 1990s. As glasnost thawed the Cold War in the early 1980s, the Eastern European communist bloc began to tolerate limited forms of political action. The legacy of Soviet central planning had caused huge levels of pollution in these countries, and the first NGOs to emerge mobilized public protest around environmental issues because it represented a relatively "safe" topic. Many of the democratic leaders who took charge in Eastern Europe after the fall of the Berlin Wall cut their teeth as environmental protesters in the 1980s, and it was this unique context that made the region highly receptive to a treaty safeguarding the rights of people to engage in environmental decision-making. UNECE seized the opportunity to turn this general political ferment into an international agreement.

Multilateral environmental agreements can take the form of declarations, which are not legally binding, or treaties, which are. Treaties include framework conventions like that signed at the Rio Earth Summit in 1992, which are broad exhortations requiring subsequent protocols, or specific laws, to be developed in order to come into force. Others, such as the Convention on International Trade in Endangered Species, which came into force in 1975, are self-standing. As Figure 4.1 shows, the number of multilateral environmental agreements has mushroomed since the 1950s, as the environment became established as an international scientific concern and environmentalism gathered momentum. The period between 1990 and 1999 stands out with 300 agreements—double the number achieved in the decades immediately before or after it. This reflects the Rio Earth Summit in 1992 and the flurry of subsequent activity generated to put various agreements into practice.

As might be expected, the proportion of multilateral environmental agreements made up by protocols and amendments has increased over

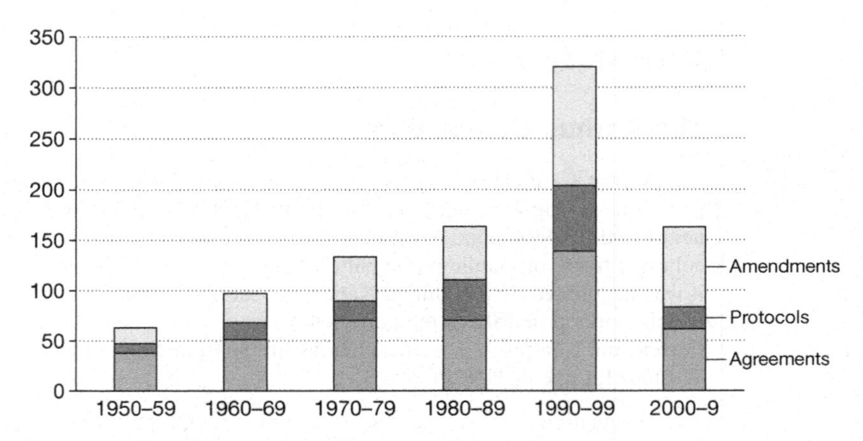

Figure 4.1 ***The growth of multilateral environmental agreements***
Source: adapted from Mitchell 2010.

time, as legal frameworks catch up with prior framework conventions, or original agreements are altered as scientific knowledge or available technologies change. While the periods 1980–89 and 2000–9 experienced similar overall numbers of multilateral environmental agreements (about 150), the latter period experienced approximately 50 percent more amendments.

Figure 4.2 gives some idea of how the focus of multilateral environmental agreements has changed over time. The proportion of agreements covering species and fish has decreased dramatically, from about 60 percent before 1960 to 30 percent in the modern era. This has been accompanied by a shift in emphasis from nature conservation to pollution control (including greenhouse gas emissions), which makes up the majority of multilateral environmental agreements after 1980. It also reflects the dominance of the systems approach in environmental science, which emphasizes the importance of habitats to maintaining biodiversity rather than focusing on individual species.

Table 4.1 lists some of the major treaties that have been agreed for different environmental threats, and their impacts. While most areas of environmental concern are now covered in some form of agreement, their success varies from effectively solving the problem in the case of ozone depletion, to having no discernible impact in the case of deforestation.

Signing is the first stage of an agreement. A treaty only comes into force when a certain number of signatories ratify, which involves translating

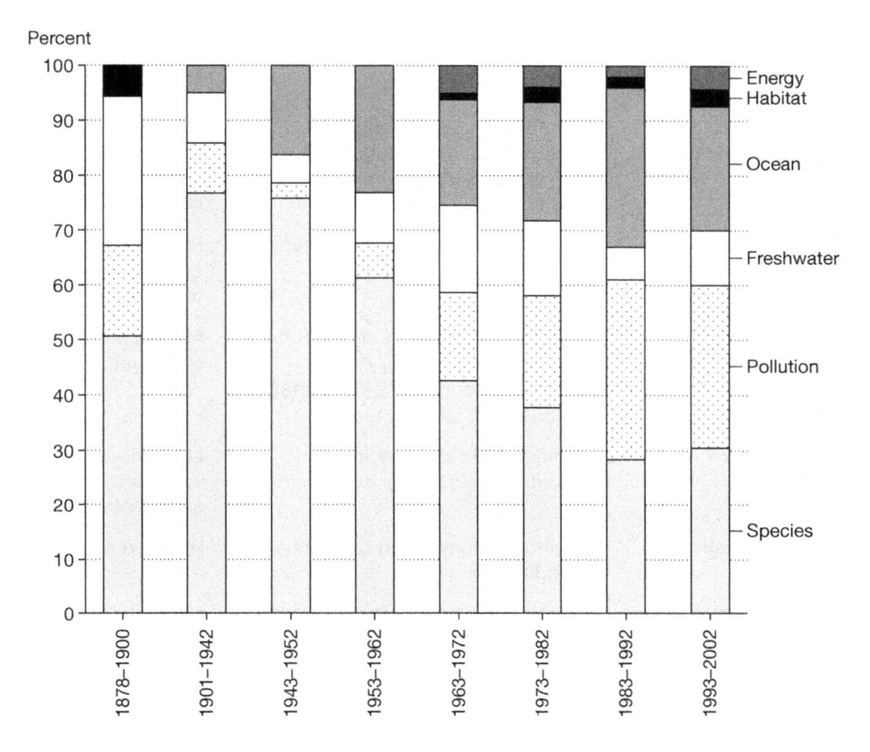

Figure 4.2 ***The focus of multilateral environmental agreements***
Source: adapted from Mitchell 2010.

the agreement into domestic law. For example, the Kyoto Protocol set out international targets for developed countries (known as Annex I countries) to reduce greenhouse gases linked to the United Nations Framework Convention on Climate Change. In order to be ratified, the Protocol required 55 Annex I nations, who jointly accounted for at least 55 percent of Annex I emissions, to secure agreement from their national legislatures (parliaments, assemblies, etc.) to implement it domestically. Once ratified by signatory countries, treaties become binding in international law (although in the absence of an international enforcement body, countries can leave a treaty at any time—in legal terms, the only thing binding them is their own consent).

Negotiating multilateral environmental agreements tends to share the problems of collective action discussed in Chapter 1. Policies tend to be diluted to a level acceptable to the least enthusiastic nation, and the free-rider effect, whereby a nation derives the benefits of others acting, say,

Table 4.1 ***Major treaty regimes***

Threat	*Treaties*	*Impact of regime*
Acid rain and regional air pollution	Convention on Long Range Transboundary Air Pollution, bilateral agreements (e.g. between the USA and Canada)	Emissions of sulfur and nitrogen dioxides now regulated, but acidified lakes slow to recover
Ozone depletion	Montreal Protocol on Substances that Deplete the Ozone Layer (1987)	Phased-out use of CFCs and effectively closed hole in ozone layer
Climate change	Framework Convention on Climate Change (signed 1992), Kyoto Protocol (signed 1997), Copenhagen Accord (2009)	Atmospheric CO_2 levels have continued to rise
Deforestation	Non-binding Forest Principles adopted at Rio (1992)	Little directly but prompted voluntary actions to certify sustainable forests
Land degradation	Non-binding Convention to Combat Desertification	Hampered by lack of funding
Freshwater pollution and shortages	Convention on non-navigable Uses of International Watercourses (not in force)	None
Marine fisheries	Convention on the Law of the Sea, whaling, plus others	Effective but not in relation to controlling over-fishing
Toxic pollutants	Basel Convention on international trade in toxic waste (came into force 1992), Stockholm Convention phasing out persistent organic pollutants, Rotterdam Convention on international trade in pesticides and industrial chemicals (both came into force 2004)	Effective among signatories
Loss of biodiversity	Convention on Biological Diversity, signed at the Rio Earth Summit in 1992	Little evidence that the convention is having an impact on species and habitat decline
Excessive nutrient loading from nitrogen-based fertilizers	None	n/a

Source: adapted from Speth and Haas 2006.

Case study 4.2

The Montreal Protocol

The Montreal Protocol, which governs the use of ozone-depleting substances like CFCs, is often held up as a model multilateral environmental agreement. Not only did the agreement phase out CFCs in a relatively short time, but it successfully involved nearly every country in the world. The agreement has also been renegotiated on several occasions, as different CFC substitutes become available and the science of ozone depletion develops. Unlike the Kyoto Protocol, the treaty also has strict trade sanctions built in to punish countries which choose to leave or break it. This mitigates the free-rider problem, as the costs to any absconding country not adhering to the terms of the agreement will outweigh the benefits they derive.

Given the difficulties of securing international agreements on other environmental issues, much ink has been spilt on the question of whether the Montreal Protocol represents a special case. The problem (CFCs used in refrigeration) and solution (CFC substitutes) were both clearly defined and supported by the commercial sector. Furthermore, the ozone hole was easily measured and the dangers posed by its depletion would affect everyone on the planet. In contrast to this, a problem like climate change is diffuse (most economic activities emit greenhouse gas), there is no simple solution supported by the commercial sector, and there will be clear winners and losers in terms of how its impacts are felt across the planet.

But some commentators have argued that ozone depletion and climate change are not so different. They point out that scientists were not certain that CFCs actually caused ozone depletion until 1988, after politicians had agreed to take comprehensive action (the Protocol was concluded in 1987). Haas (1992) suggests that the political will to act even in the face of scientific uncertainty can be explained by the existence of a strong international community of scientists in organizations like UNEP and the US Environmental Protection Agency, who were pivotal in persuading the US government, the largest consumer of CFCs, and DuPont, the largest global manufacturer of CFCs, to support their phasing out plans. This network of experts formed an "epistemic community," who shared a common understanding and proposed solution to the particular problem of CFCs (Bulkeley 2005). They targeted key actors so that other countries and companies would follow, and shaped the decision-making process by determining what alternatives were on offer.

One of the reasons the Montreal Protocol worked was that US industry was ahead of its competitors in developing CFC substitutes. Initial concern about the ozone layer in the 1970s led to a ban on the use of CFCs as aerosol propellants (although not as refrigerants) in several countries, including the USA. Many European countries, under pressure from industry, did nothing. While US companies resented the competitive advantage that their European rivals had achieved, this meant that they were in a stronger position when the Montreal Protocol was under discussion as they had already developed alternatives. Although critics point out that DuPont actually stopped research on CFC alternatives in 1981, the $600 million they stood to lose per year through the complete phasing out of CFCs comprised only 2 percent of their overall income, allowing them to take a longer-term view based on innovation.

This analysis is intriguing for many reasons. The US government and large corporate interests are usually cast as the villains of global environmental governance, but were central players in this case. Networks of experts also played a critical role in exerting behind-the-scenes influence over key players to create a regime that was willing to change (Litfin 1994, Young 2008). The recent effort by leading climate scientists to establish a 2°C guardrail, which policy-makers must not allow to be broken, constitutes a similar attempt to establish a strong epistemic community to drive action on climate change. Whether this particular epistemic community succeeds is yet to be seen.

to reduce atmospheric pollution, without having to take action themselves, remains. The sheer number of countries involved in international meetings can make it exceptionally hard to reach legally binding agreements. Case study 4.2 discusses how these problems were circumvented by one multilateral environmental agreement, the Montreal Protocol on Substances that Deplete the Ozone Layer. The Montreal Protocol is often held up as a model of international cooperation, and it offers insights into the role of science in securing environmental agreements that have relevance to current efforts to secure a legally binding climate agreement.

The United Nations conferences

Since its inception in 1945, the UN has facilitated the establishment of an international regime around environmental issues. Meetings such as the Scientific Conference on Conservation and Utilization of Natural Resources in 1949 broadly addressed issues that we recognize today as environmental, but it was with the Conference on the Human Environment in Stockholm in 1972 that these concerns moved from the realm of science into the political arena. Barbara Ward, a British economist, and René Dubos, a French-American microbiologist, were commissioned to write a report that would form the conceptual basis for discussions in Stockholm. The ensuing document, *Only One Earth*, drew strongly on the one-world discourse to advocate "loyalty to the Earth."

As well as showing that there is nothing that novel about enlisting respected economists and scientists to raise environmental awareness among politicians, the report laid the groundwork for the concept of sustainable development—Ward is generally credited with coining the phrase "spaceship Earth," and Dubos the phrase "think global, act local."

Stockholm was the first time that environmental concerns had been explicitly linked to the need for development in poorer countries. When the Indian prime minister, Indira Gandhi, addressed the conference, she supported such a union, asking, "Are not poverty and need the greatest polluters?" The argument that lifting people out of poverty will enable them to better protect their environment formed an attractive storyline for the UN, many of whose members were in the developing world and more concerned with development than the environment. The principle of additionality was outlined at Stockholm to address these concerns, whereby the developed world must help pay for environmental protection in the less developed world.

But the equation between development and environment sits uneasily with many developing countries. The idea that affluence breeds environmental concern is well established—environmentalism was born in the developed world as people recoiled against the damage wrought by industrialization and urbanization and became able and willing to pay to prevent it. While research has shown that communities in the developing world are equally as concerned about environmental issues (Dunlap and York 2008), their politicians remain wary of Western environmentalism, which they see as a threat to their economic development. While these difficulties were to bubble away beneath the surface, Stockholm has generally been portrayed as a success in setting an international environmental agenda for the first time.

In 1987, the World Commission on Environment and Development (WCED), chaired by the prime minister of Norway, Mrs. Gro Harlem Brundtland, published the report *Our Common Future*, which introduced the concept of sustainable development to the world (World Commission on Environment and Development 1987). Famously defined as "development which meets the needs of the present without compromising the ability of future generations to meet their own needs" (ibid.: 43), sustainability rests upon the assumption that economic growth, social well-being and environmental protection can be organized in such a way as to be mutually supporting. It sought to square the circle between environment and development by showing how economic growth could be decoupled, or separated, from its negative environmental impacts.

As discussed in Chapter 2, sustainable development emerged at a similar political moment to governance. In the context of rapid economic globalization in the late 1980s, the idea that economic growth could be allied with environmental protection provided a way to unite leaders in both the developed and developing worlds behind a common goal. Further, in making links between economic, social and environmental policy, sustainable development presents a set of challenges around integration and coordination to which governance is suited (Kemp *et al.* 2005).

The United Nations Conference on Environment and Development, held at Rio in 1992, was designed to take stock 20 years after Stockholm. Billed rather grandiosely as the "Earth Summit," it was a massive event, attended by 153 countries, 2,500 NGOs, 8,000 accredited journalists, and an estimated 30,000 hangers on. Tasked to mainstream sustainable development into national policy, a number of agreements were signed at Rio, including the Framework Convention on Climate Change, the Convention on Biodiversity, Agenda 21, the Rio Declaration, the Forest Principles, and the Convention to Combat Desertification. The principle of subsidiarity, which "states that decisions within a political system should be taken at the lowest level consistent with effective action" (Jordan and Jeppesen 2000: 66), was also enshrined in Agenda 21.

Rio turned sustainable development into a household word, but, lacking financial and legal commitments (especially to forests), some considered its achievements less than impressive. Tensions between the developed and developing worlds continued to simmer, with developing countries shying away from a legally binding forest convention, which they feared would cede control of their forests to rich countries. Jonathon Porritt, one of the founders of the UK Green Party, despondently reported, "I came here with low expectations, and all of them have been met" (Diamond 1992). The Zairese (now the Democratic Republic of Congo) delegate bluntly stated, "If this kind of Earth Summit circus continues, then the people of Africa will perish. We need the rule of law. We need democracy, peace with justice, and we need fair terms of trade so we can develop a proper market economy, then we can protect our environment" (quoted in Jordan and Voisey 1998: 94). As Banerjee (2008: 65) has noted, "Slogans, however pretty, do not make a theory." While the Earth Summits rest upon the dual assumptions that the environment is a global issue, and that there is a global community capable of governing it, friction between the developed and developing worlds suggests otherwise. Some of the key tensions surrounding the "one world" discourse are discussed in Key debate 4.1.

In 1997, the UN held a General Assembly Special Session on Sustainable Development in New York to review progress five years after the Rio Earth Summit. This meeting highlighted the huge cost of sustainable development and the related failure to implement many of the agreements and accords that were signed. The meeting concluded that a major international follow-up to Rio was required to kick-start progress on implementation. It was in this context that the UN organized a second Earth Summit, held in Johannesburg in 2002, to focus on concrete actions capable of implementing the words and aspirations of Rio. Attended by 10,000 delegates, 8,000 representatives of major groups and 4,000 members of the media, Johannesburg more than matched the scale of Rio (United Nations 2002). But unlike the previous Earth Summit, Jo'burg focused more on the role of society, shifting the agenda from the science of environmental change to the question of how to implement sustainable development.

Unfortunately, this meant that the conference ran up against a set of wider and more intractable political and economic realities (Speth and Haas 2006). The problems were all too clear. Only 10 percent of $125 billion in aid that flows from the developed to the developing world is directed at the basic needs of the poorest countries, while rich countries persist in subsidizing industries that pollute the global environment and prevent those in the developing world from competing in a fair market place. Agriculture is often singled out as the main culprit, being subsidized annually to the tune of $15 billion in the USA and €48 billion in the EU. These so-called "perverse subsidies" perpetuate a system that is over-reliant on fossil fuel inputs for machinery and fertilizers, destroys biodiversity, and makes it impossible for farmers from poorer countries to obtain fair prices on the world market for products that they can easily grow (Myers and Golubiewski 2007). Although much hyped, there was a lack of political will to reach agreements on targets for emissions, while fairer terms of trade for the developing world were hampered by the fact that such matters were deferred to the WTO, which was itself failing to make headway on terms of trade between the developed and developing worlds.

The main outcome of the meeting was the Johannesburg Declaration on Sustainable Development, but this was little more than a re-affirmation of previous commitments made at Stockholm and Rio. In this sense

Key debate 4.1

One world, one Earth?

The notion that the environment is not only governable but best governed at the global level is largely taken for granted today. As Jasanoff notes (2004: 32), "The idea that there is 'only one earth' seems to have lost its sloganeering quality and been accepted as reality by activists and policy-makers, the media and the public," exerting a huge influence over the way in which environmental problems are addressed. But the idea of the global environment is predicated upon the assumption that there is a global "we" to care about it. Given that the world is made up of myriad different peoples and cultures, it is hard to identify who this global "we" really is. In the original *Our Common Future* report, the global "we" is based on nothing more than the fact that humanity occupies a single planet. One does not have to look far to find examples that suggest spatial proximity does not automatically engender unity. The Middle East springs readily to mind.

The one world discourse also implies an undifferentiated response to environmental problems, whereby a common problem is taken to infer that there is a common solution. This has two adverse effects. First, it turns people into passive spectators, waiting for solutions to be passed down from on high, rather than acting themselves. Second, the vision of one world is place-less, smoothing over local circumstances and paying little attention to the views of people that do not fit with it. The effects of this range from resentment on the part of developing countries who feel forced to take environmental action to address problems that have been created by the developed world, to the exclusion of indigenous rainforest tribes from international discussions surrounding their future (Fogel 2004). This problem is exacerbated by the fact that global institutions are dominated by developed countries, which tends to marginalize the developing world in international environmental negotiations. The irony is that many issues, like deforestation and biodiversity, will affect exactly those countries that are excluded or sidelined (Agrawal *et al.* 1999).

The one world ideal is a *sine qua non* of global environmental governance, substituting for the existence of a formal global polity. Within the faceless humanity of *Our Common Future* people become interchangeable units in environmental policy, with those who do not fit the template of the global citizen sidelined. As Fues *et al.* (2005: 243) state, to avoid overlooking the interests of different nations in the name of the common good, "conflicting interests have to be precisely named and not hidden behind an idealized superior common interest."

the World Summit on Sustainable Development, damningly dubbed "Rio minus 10," failed to achieve its goals. But worse, it highlighted a deeper tension, that richer countries would at some point have to make economic sacrifices in order to address global environmental issues in an inclusive way.

While the UN conferences are often criticized for producing more heat than light, it is important to recognize what these meetings have achieved. The environment is relatively well advanced as an international issue when compared to other global concerns, and it has assumed this level of prominence in little over 40 years (Fairbrass and Jordan 2005). Seyfang and Jordan (2002) identify six positive functions of what they call "environmental mega-conferences" like Rio and Jo'burg:

Setting global agendas: establishing specific issues as being of international importance.

Facilitating joined-up thinking: showing how environmental issues relate to economic, political, and social questions.

Endorsing common principles: forging shared understanding between nations and people.

Providing global leadership: offering a focus around which countries can coalesce.

Building institutional capacity: creating organizations that are capable of coordinating international action.

Legitimizing global governance: widening involvement at, for example, Jo'burg, to ordinary people and a wide range of NGOs.

As Seyfang states (2003: 227), "Environmental mega-conferences do serve an important function in contemporary environmental governance, even though they are not the panacea that some had originally hoped they might be." Given the structural constraints on how global environmental governance takes place, the achievements in the environmental field become rather impressive, if still falling short of what we might desire in an ideal world. Securing *any* form of agreement between almost 200 countries, which have vastly differing agendas and are driven by multiple tensions, through only the power of persuasion is nothing short of a miracle. The story, to date, is perhaps one of increasingly shared understanding, if not action.

The shift to climate change

From 1992 onwards the focus of global environmental governance has shifted to one problem in particular—climate change. High-profile global threats like droughts, flooding and extreme weather events have propelled climate change up the political agenda, and although still addressed within the broad framework of sustainable development, climate change has prompted its own series of high-profile international meetings. The question of how to make development sustainable has simply narrowed to focus on how to decouple economic growth from greenhouse gas emissions.

The United Nations Framework Convention on Climate Change (UNFCCC) was one of the multilateral environmental agreements signed at the Rio Earth Summit in 1992. The treaty aims to control the emission of greenhouse gases in order to prevent atmospheric warming and the negative consequences associated with it. While the treaty itself was not legally binding, it set out a roadmap for subsequent protocols to limit greenhouse gas emissions and establish enforcement mechanisms. A total of 192 parties signed the treaty (with the high-profile exception of the USA), and annual Conferences of the Parties (CoPs) have been held from 1995 onwards to assess progress in dealing with climate change. These led to the Kyoto Protocol, which established legally binding obligations for developed countries to reduce their greenhouse gas emissions. Although the Protocol was adopted in 1997, it only came into force in 2005 when the ratification criteria had been met.

Parties to the United Nations Framework Convention on Climate Change are divided into Annex I countries, which include 39 industrialized countries and the EU, and Annex II countries, which are a subset of Annex I comprising the OECD members who were not "economies in transition" (i.e. post-Soviet countries) in 1992. Annex II countries are also committed to paying for the costs of developing countries making emissions reductions. While all member countries have made a general commitment to reducing greenhouse gas emissions, Annex I countries are committed to reducing four greenhouse gases (carbon dioxide, methane, nitrous oxide, and sulfur hexafluoride), and two groups of gases (hydro-fluorocarbons and per-fluorocarbons) produced by them, to 5.2 percent below their 1990 level (Grubb *et al.* 1999).

Benchmark emission levels for 1990 were calculated using figures from the IPCC Second Assessment Report, and the emissions of various greenhouse gases were converted into CO_2 equivalents. Emissions from international aviation and shipping are excluded from the targets, as are industrial gases like CFCs that are dealt with under the Montreal Protocol. Given the economic crisis that befell most economies in transition in the 1990s, the economic output and emissions of these countries remains well below 1990 levels and they are not obliged to act.

The Kyoto Protocol set up a number of flexible mechanisms to allow Annex I countries to meet their targets. Emitters were allocated a certain number of emissions credits based upon their need, which they were then allowed to sell or purchase from elsewhere in order to meet targets. This could take the form of funding emissions reduction projects in non-Annex I countries (the clean development mechanism), or by simply buying and selling excess credits from other Annex I countries. Under the Kyoto Protocol, developing countries are not required to reduce emissions levels as it would hamper their economic development. They are allowed to sell emissions credits to Annex I countries, which can be generated by projects to remove carbon from the atmosphere (most commonly reforestation), and they get funding and technology for low-carbon development.

The Kyoto Protocol expires in 2012, and, with it, the commitments of member countries to report and reduce their greenhouse gas emissions. It was this timeframe that lent such urgency to the negotiations at the Copenhagen Climate Conference in December 2009. Although the conference was simply the 15th annual CoP meeting, it assumed greater importance as a symbol of global commitment to tackle climate change. Negotiations failed to produce a legally binding agreement though, with members only signing a last-minute accord that did little more than indicate their acceptance that something must be done. While this was heralded as a calamitous failure by the world's environmental lobby, others have hailed Copenhagen as an important step in establishing a global commitment. While Kyoto was legally binding, not many countries were actually bound. In the aftermath of Copenhagen all major emitting countries have committed to climate change.

The major barrier to achieving a legally binding commitment to reduce greenhouse gas emissions no longer concerns the scientific evidence surrounding warming, but the question of who should reduce emissions and by how much (Bohringer 2003, Najam *et al.* 2003, Rose 1998). Given that there is a direct correlation between economic activity and carbon emissions, most countries worry about the financial costs of reducing emissions—indeed this was the reason that George Bush gave for not ratifying the Kyoto Protocol.

The split between Annexed countries and the developing world is intended to recognize that the latter cannot afford the costs of emissions reductions. But critics from the developed world argue that both developing countries and developed countries need to reduce their emissions to address climate change, regardless of what is fair, otherwise increased emissions associated with the huge levels of economic and population growth in developing countries will vastly outweigh any reductions achieved by developed countries. The Byrd–Hagel Resolution, which states that the USA will not sign any emissions reduction protocol unless it includes developing countries, was passed in 1997 in the US Senate by 95 to 0 (Helm 2000). Commentators from the developing world see such demands as simply extending the legacy of Western colonialism, which continues to repress the development of the poorest countries in order to maintain the dominance of the West (Agrawal 1995, Agrawal and Narain 1990). As Figure 4.3 shows, the overall emissions of greenhouse gases remain heavily concentrated in the developed world.

The developing world argues that rich industrialized countries have been historically responsible for most emissions, because they should bear the cost of fixing the problem. The use of 1990 emissions levels as a benchmark is intended as something of a compromise between these two positions, as this is taken to be the point at which the threat of climate change became widely accepted and hence only emissions after this point can be deemed irresponsible. These tensions have actually become enshrined in the subtly different definitions of climate change that are at play in global environmental governance, discussed in Key debate 4.2.

The fairness of the emissions benchmarking system used in the Kyoto Protocol has also been questioned, as benchmarking penalizes countries that have already made efforts to reduce their emissions and rewards those who have not (Goldemberg *et al.* 1996). For example, two countries might have had identical levels of emissions in 1990, but one may have made significant prior efforts to reduce their emissions, while the other had not. Although both will receive the same reduction targets,

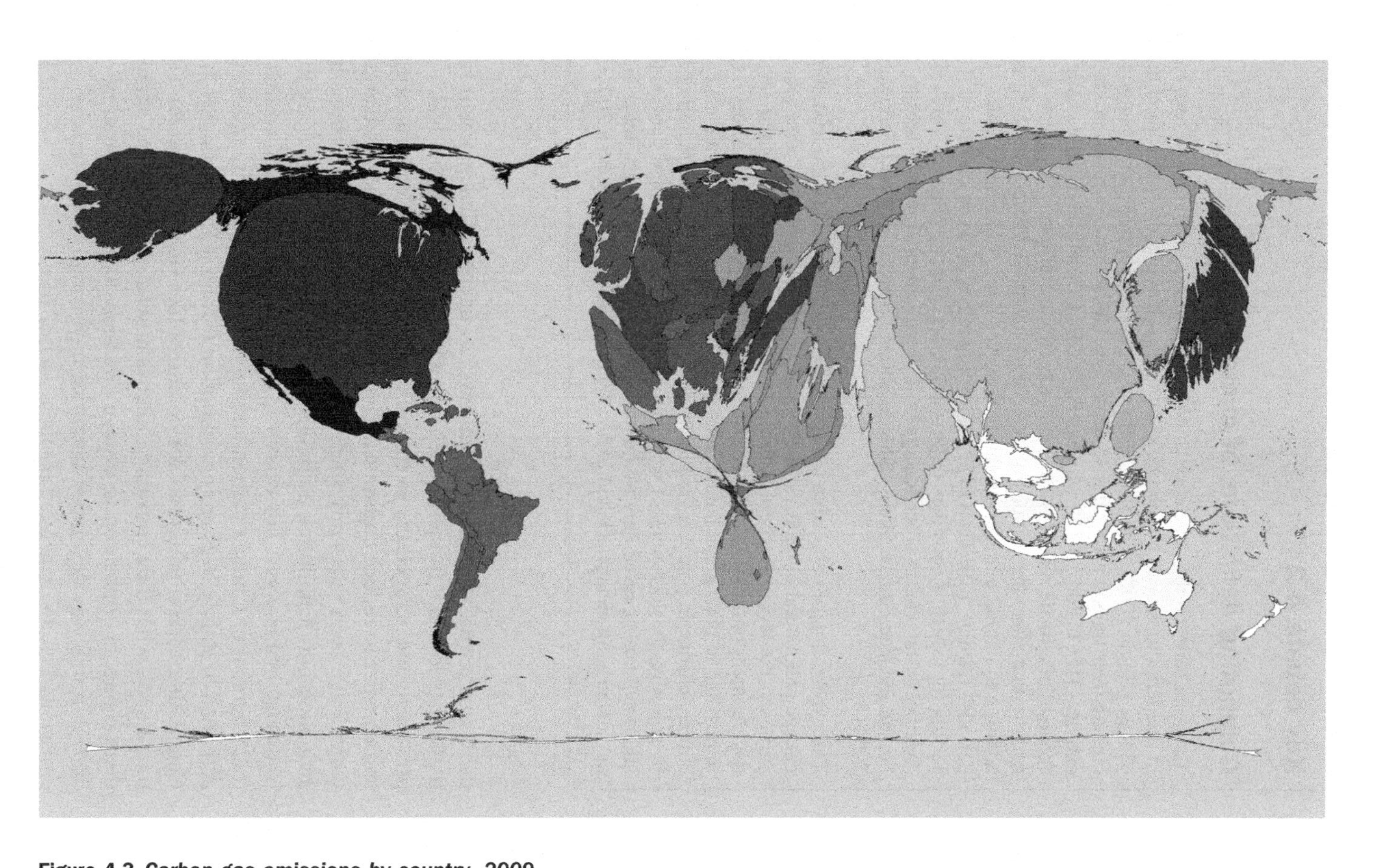

Figure 4.3 ***Carbon gas emissions by country, 2009***

Source: www.views of the world.net. Copyright SASI Group (University of Sheffield), reproduced courtesy of Benjamin Hennig.

Key debate 4.2

Competing definitions of climate change

Tensions between the developed and developing worlds even find expression in the definitions used by the two central organizations dealing with climate change (Uggla 2008). The IPCC defines climate change as "any change in climate over time, whether due to natural variability or as a result of human activity" (IPCC 2007: 871), whereas the UNFCCC defines climate change as "change of climate which is attributed directly or indirectly to human activity that alters the composition of the global atmosphere and which is in addition to natural climate variability observed over comparable time periods" (Article 1.2). The UNFCCC distinguishes between climate variability, which it considers to be natural, and climate change, which is specifically human, in order to suggest that only adaptation measures to human-induced climate change should obtain financial support (Verheyen 2002).

The expectation of being able to distinguish between human-induced climate change and natural climate variability reflects the reluctance of annexed parties to the Kyoto Protocol to provide financial support for regular development projects. However, the formulation is problematic, since it is impossible to distinguish natural climate variability and human-induced climate change in practice. Instead, expectations of such a distinction result in awkward considerations of what can be defined as additional harm and additional costs caused only by human-induced climate change, rather than any underlying changes (Klein *et al.* 2003, Pielke 2005, Verheyen 2002).

the country with an already decent level of energy efficiency will face high costs to make further reductions, while the country that had previously encouraged the overconsumption of energy will find it easier and cheaper to reduce emissions. That said, countries such as Germany, which have been massively successful in reducing their emissions by switching out of polluting activities, still import products that are made in unsustainable ways. For example, recent studies have shown that up to 25 percent of China's carbon emissions are generated by industrial activities to satisfy Western consumption (Wang and Watson 2007). Calculating carbon emissions for countries like Germany and the UK based upon what they consume rather than what they produce shows them rising, in opposition to official emissions figures (Helm *et al.* 2007). In this way developed countries have simply exported the emissions that are generated by their way of life.

Finally, there is the familiar realist problem of countries simply acting in their own interests, rather than collectively. Reports suggest that the failure of the Copenhagen Climate Conference to reach anything approaching a legally binding agreement was largely due to the efforts of China, in cahoots with the Sudanese and Ethiopian delegations, to derail the negotiation process. For example, they objected to the usual process whereby a small group of representative countries meet and report back to the wider assembly, which led to unwieldy discussions between huge numbers of representatives. They also sent different negotiators to different sessions, with premier Wen Jiabao remaining absent from the final discussions over the accord. Even in agreeing to the accord they demanded the removal of any binding targets for reductions, including both the target for industrialized countries to achieve an 80 percent reduction in emissions by 2050, and the overall target of a 50 percent reduction by 2050.

Interestingly, with support from India, Brazil and South Africa, China also refused to commit to developing legally binding protocols for emissions reductions in the future. As discussed above, the developing world is wary of any legally binding agreements on emissions that may hamper their economic growth, and China's alliance with Sudan, leader of the G77 that represents the developing world countries, was highly strategic. Copenhagen produced a frustrating outcome, in that previous agreements were not only re-hashed, but actually watered down.

The paradox of Copenhagen is that China has been taking substantial actions to reduce its emissions by 40–45 percent, putting Western governments to shame with its levels of investment in renewable energy (Browne 2010). The back story to Copenhagen has more to do with the battle for global supremacy between the USA and China, in which neither wants its hands tied by binding agreements, than the actual Chinese commitment to addressing climate change. It is here that global governance runs headlong into the realities of international relations, whereby countries compete for individual advantage.

Architecture

The architecture of global environmental governance is made up of institutions that are created to enact the agreements reached at international meetings. Arguably the most important environmental institution, UNEP, was created to implement the mandates of the

Stockholm Conference on Human Development in 1972, while the Commission for Sustainable Development was set up to review progress on the Rio agenda from the 1992 Earth Summit. Each major treaty has Conferences of the Parties (recognizable in the titles of interim meetings by its acronym, CoP), who may meet regularly and are serviced by either UNEP or their own secretariat.

This can produce a rather fragmented institutional landscape. So, for example, while the Basel Convention on international trade in toxic waste (1992), the Stockholm Convention on persistent organic pollutants and the Rotterdam Convention on international trade in pesticides and industrial chemicals (both 2004) all deal with hazardous materials, they are administered by separate secretariats. This actually led to an extraordinary CoP meeting, which brought the three secretariats together to rationalize their work and make fewer demands on member countries. Such synergies promise a potential way to unify governance institutions from the bottom up: for example, it has been suggested that the Montreal Protocol could also be subsumed within this joint CoP, as CFCs are also a hazardous material.

There is no single international body with powers to develop and enforce environmental policy, which represents something of a problem. The power of existing global bodies like the UN and the WTO should not be overstated, as they are still funded by and thus beholden to their member states. The contrasts between the EU and the UN are instructive here. The EU Commission sets a high bar for policy which the nation states bring down to achievable levels, while the UN Secretariat has to respond to the competing demands of over 190 member countries, and cannot impose its own strategy or policy vision. As a result, although the environment is framed as a global problem, "it is precisely at this level that government institutions are least effective and trust most delicate" (Levin *et al.* 1998: 233). The relative weakness of global environmental institutions to enforce action stands in stark contrast to the generally accepted credo that environmental issues require global action. Unsurprisingly, this has led to calls to create a more powerful "world environmental organization," discussed in Key debate 4.3.

Implementation

Implementation is the most important but least glamorous element of global environmental governance. It needs most money but usually has

Key debate 4.3

Do we need a world environmental organization?

Three broad models for a world environmental organization can be distilled from the considerable literature on the subject (Biermann 2001, Lodefalk and Whalley 2002):

Cooperation Upgrade UNEP into a specialized agency like the World Health Organization.
Hierarchization Create an agency with executive decision-making and enforcement powers.
Centralization Streamline and integrate existing agencies, programs and initiatives.

A common proposal suggests forming an institution akin to the WTO, which has had success in integrating trade agreements and opening up markets because it is able to apply legal pressure on nation states and resolve disputes (Biermann 2005). But environmental problems are very different to trade disputes. Markets are socially constructed with rules that can be negotiated and renegotiated, whereas aspects of the environment such as the hole in the ozone layer are not (Najam 2003). Further, one of the strengths of current environmental governance is that it is broadly inclusive of NGOs and civil society. This stands in stark contrast to the accusations of introversion leveled at international organizations like the WTO, which are seen by many to be dominated by narrow corporate interests. Major public protests against the WTO suggest that there is less than complete public support for its activities. Power rests on legitimacy, and, even if it existed, no such organization could realistically enforce unpopular environmental measures at the global level.

The idea of centralizing governance functions also threatens to undermine some of the more effective aspects of contemporary global environmental governance (Najam 2003). With fragmentation comes flexibility, and the diverse set of institutions currently addressing environmental issues allows them to respond more effectively and forge links across different domains. Similar criticisms can be made of current suggestions to form a global climate bank (German Advisory Council on Global Change 2009). Following this line of argument, Oberthür and Gehring (2004) suggest that the creation of a world environmental organization would offer little more than institutional restructuring for its own sake. Addressing climate change effectively is not simply a matter of re-arranging the administrative chairs on the *Titanic*, but of addressing issues of global justice and unfair terms of trade. While the debate over the global institutional framework for environmental issues will undoubtedly rumble on, there is currently little support for any one proposal.

least, and is beset by disputes concerning who should provide funds and under what conditions. Funding sources for global environmental initiatives include nations, groups of nations like the EU, the UN itself and international financial institutions like the World Bank and Inter-American Development Bank. Traditionally funding takes the form of low interest loans, but increasingly NGO and private money is being used as well.

Most funds are channeled through the Global Environmental Facility (GEF), established in 1991 as a $1 billion multilateral environmental funding mechanism in the World Bank to protect the global environment and promote sustainable development. Based on the principle of additionality, the GEF covers costs associated with transforming a project with national benefits into one with global environmental benefits. The three initial partners were the United Nations Development Programme, the United Nations Environment Programme and the World Bank. In 1994 the GEF was restructured and moved out of the World Bank system to become a permanent, separate institution in order to enhance its legitimacy with developing countries, which have traditionally been suspicious of the neoliberal leanings of the World Bank. Simultaneously, the GEF was entrusted to become the financial mechanism for both the UN Convention on Biological Diversity and the UN Framework Convention on Climate Change.

Today, the GEF is the largest funder of projects to improve the global environment, allocating some $8.8 billion, supplemented by more than $38.7 billion in co-financing, for in excess of 2,400 projects in over 165 developing countries and countries with economies in transition. While impressive, these totals fall far short of what is required. Official development assistance stands at around $50 billion per year globally, yet estimates suggest that the agreements signed at the Rio Earth Summit alone require $200–500 billion per year to implement (Saunier and Meganck 2009).

One of the major achievements at the Copenhagen Climate Conference was to establish a fast-track finance scheme for clean development and adaptation. Governments committed $30 billion for the period 2010–20, and agreed to establish a long-term fund of $100 billion by 2020. These figures are relatively large, given that the GEF recently struggled to raise $3 billion, and the International Monetary Fund is considering how to mobilize this fund.

Case study 4.3

The Millennium Development Goals

At the UN Millennium Conference in 2000, 147 states adopted the Millennium Development Goals (MDGs), which provide eight targets to reduce poverty by 2015. The goals represent perhaps the most far-ranging attempt to coordinate global action on sustainability, and include eradicating extreme poverty and hunger, promoting universal education, improving gender equality, improving child and maternal health, combating HIV/AIDS, ensuring environmental sustainability, and enhancing global partnerships. Then secretary-general, Kofi Annan, established the three-year Millennium Development Project to identify country-specific measures that might be taken to achieve the goals, and set up taskforces for each goal to work with countries to figure out how best to address problems like reducing child mortality rates. Although housed within the UN Development Programme, the delivery of the Millennium Development Goals involves numerous UN bodies, and plays an important role coordinating the UN machinery in its delivery of sustainable development.

While it is becoming apparent that the targets are not going to be met globally, there is great variation between different regions. So, East Asia is set to meet targets related to poverty, while much of sub-Saharan Africa definitely will not. Almost nowhere is expected to make the targets for environmental sustainability. From a governance perspective, initiatives like the Millennium Development Goals coordinate global efforts to address development and the environment, but are hobbled by lack of funds and inadequate institutional capacity in specific countries. For example, 0.7 percent of donor country GDP would equate to about $200 million, compared to the current donor flow of about $70 million, enough to deliver the goals in well governed countries according to the UN Millennium Project. While this has been promised, it has not been delivered (Japan and the UK are the only countries to honor their development aid commitments).

Unfortunately, many of the poorest countries have inadequate institutional infrastructure to implement initiatives even if the funding were available. This also makes measuring progress towards the goals tough. For example, while many organizations monitor incomes, only one African country (Mauritius) records basic events such as births and deaths in accordance with UN guidelines, and details concerning costs and what happens in the intervening period between birth and death are largely a matter of guesswork (Attaran 2005). Implementation is thus hampered by a lack of monitoring, which makes progress difficult to measure.

Not only is implementation the most expensive part of global environmental governance, but it is also the least appealing. While the benefits to politicians of attending high-profile conferences are obvious, implementation involves the kind of long-term commitment to action that requires significant amounts of resources and rarely produces headlines. Further, the sheer number of meetings and resulting secretariats has produced a complex institutional terrain, fragmenting implementation across multiple organizations and hampering the ability of departments in member states to cooperate. The density of regimes surrounding different but related environmental issues leads to incoherence and a resulting lack of implementation. This has led to accusations that there is more talk than action at the global level, with sustainable development singled out for criticism as "the mantra that launched a thousand meetings."

Perhaps most dangerously, the perceived lack of implementation has also driven a growing skepticism among the international community concerning the worth of such treaties, who want evidence that they are having a positive effect. Western governments have become increasingly conservative in recent years, while Eastern governments are showing signs of fatigue and indifference to large-scale treaty negotiations. Some of the problems of implementation are discussed in Case study 4.3, which looks at the UN's Millennium Development Goals.

Conclusions

The global environmental agenda has evolved through a series of high-profile meetings organized by the United Nations, which have produced a number of agreements between nations to address specific environmental problems. The story can be explained as one of excitement and agenda setting in the early days, followed by an increasing recognition of the need to implement agreements and secure legally binding commitments.

Table 4.2 lists the key challenges for each element of global environmental governance. Overall, the process is primarily reactive, and conflict between states means that negotiations are often lengthy and result in few legally binding agreements. This translates into a fragmented institutional landscape, with secretariats representing different sets of signatory countries for each agreement, and UN bodies sharing responsibility for overlapping policy issues. In turn, this hampers

Table 4.2 ***Key challenges for global environmental governance***

Process	*Architecture*	*Implementation*
Reactive	Fragmented institutional landscape	Hard to coordinate and fund action
Conflict between states	Secretariats have incomplete and overlapping memberships	Lack of unified action
Few legally binding agreements	Lack of authority	Can't force action
Lengthy nature of negotiations	Plethora of institutions	Large resource cost of negotiating

the coordination and funding of concerted action around specific environmental problems. There is little doubt that the procession of huge, high-profile meetings has generated massive interest around environmental issues and established them as part of the international agenda, but their failure to provide solid grounds for progress shows frustratingly few signs of improving.

Looking at the bigger picture, Park *et al.* (2008) identify two key failings of the system of global environmental governance:

Underestimation of economic forces. The current dominance of international financial flows and economic growth was largely unforeseen when the current global institutional architecture was put in place, and has changed the parameters within which global action can be taken. In the post-war period, there has been a gradual shift of international power from the UN to global financial institutions like the World Bank and the WTO. For example, imposing levies on unsustainable imports is effectively a legal question for the WTO, not UNEP.

Focus on trans-boundary issues at the expense of global systems. Early successes, like the prevention of acid rain that involved only a few countries and the restriction of CFCs to protect of the ozone layer that involved only a few companies, were far simpler physically and politically than the environmental problems faced today.

Ironically, the agreements produced by the current system of global environmental governance prioritize state action. For example, the Brundtland Principles all begin with the words "States shall . . . " As liberalist international relations scholars emphasize, it is non-state actors who constitute global governance. For them, the words "international" or "intergovernmental" limit the game to nation states, whereas global governance is actually enacted by various organs of civil society like NGOs. While commentators like Speth and Haas (2006) suggest that weak treaties are to blame rather than weak implementation, the idea that we need stronger global government should be viewed warily. Given that it is through civil society networks that implementation is most likely to take place, an effective governance system needs to be both decentralized and flexible. If this is indeed the case then it is perhaps premature to jettison governance in favor of some monolithic state-sponsored global body.

The foreseeable future will involve working with what we have, which means harnessing networks and markets to address the complex economic and political challenges identified by Park *et al.* (2008). It is with this in mind that we turn to explore the role of networks in environmental governance.

Questions

- Where does international cooperation end and global governance begin?
- Do you agree with Bill McKibben that "environmentalists have failed to make measurable progress on the greatest challenge anyone's ever faced . . . So we better come up with something new"?

Key readings

- Biermann, F. and Pattberg, P. (2008) "Global environmental governance: taking stock, moving forward," *Annual Review of Environment and Resources*, 33: 277–94.
- Fues, T., Messner, D. and Scholz, I. (2005) "Global environmental governance from a North–South perspective," in A. Rechkemmer (ed.) *UNEO: Towards an International Environment Organization*, Baden-Baden: Nomos, 241–63.
- Saunier, R. and Meganck, R. (2009) *Dictionary and Introduction to Global Environmental Governance*, London: Earthscan, chapter 1.

Links

- http://environmentalgovernance.org/. Home to the Global Environmental Governance Project, a joint initiative hosted by the College of William and Mary and the Yale Center for Environmental Law and Policy, which aims to focus academic effort in order to strengthen environmental policy-making at the global level. Excellent set of videos.
- http://iea.uoregon.edu/. The International Environmental Agreements website, hosted by the University of Oregon, which provides a storehouse of information concerning environmental agreements, categorized by type, subject, date, membership and so on.
- http://itunes.apple.com/us/app/negotiator/id338997029?mt=8. UNFCCC ITunes application providing quick and easy access to essential information about the UN Climate Change Conferences taking place and to allow virtual participation.

5 Networks

Intended learning outcomes

At the end of this chapter you will be able to:

- **Understand the power of networks to coordinate environmental action.**
- **Evaluate the characteristics and importance of transnational governance networks.**
- **Appreciate the role of voluntary networks in making business more sustainable.**
- **Assess the strengths and weaknesses of network governance.**

Introduction

> If there is no wind, row.
>
> (Latin proverb)

Governance networks bring civil society and private organizations together voluntarily to address environmental issues (Bäckstrand 2008). Within network governance, groups of stakeholders with vested interests in a decision form self-organizing networks that work together to achieve common goals and mutually beneficial outcomes (Rhodes 1996, Rhodes and Marsh 1992). Networks are seen as critical in implementing multilateral environmental agreements, because they utilize the existing resources of multiple actors, and avoid the impasses of multilateral action by simply bypassing reluctant national governments.

This chapter begins by outlining the characteristics of networks that lend them power, and considers how networks can be managed and analyzed. It then looks at the role of transnational networks in governance, exploring the renewable energy network REN21 in depth. Corporate social responsibility is discussed in relation to voluntary certification and auditing schemes, which enroll businesses into environmentally friendly actions. These are explored using the example of the Forest Stewardship Council's certification scheme for sustainable timber products. The chapter concludes by assessing the strengths and weaknesses of network governance.

The power of networks

Networks are emblematic of the shift from government to governance, whereby multiple independent actors are linked by voluntary rather than legal agreements (Jones *et al.* 1997). Following Klijn and Skelcher (2007), "network governance" is taken to mean the broader way in which society and politics are organized (i.e. the mode of governance), while "governance networks" are the actual units of governance. Being voluntary in nature enables networks to govern themselves, which allows them to be more responsive to emerging needs and opportunities than either state bureaucracies or markets, which operate within legally constrained regulatory frameworks. Networks can also grow quickly by enrolling new members, pooling resources to achieve things that would be impossible for its constituent organizations working alone. Stakeholders are bound together by the belief that they have complementary strengths which allow them to achieve shared goals more effectively if they work together. This so-called "capacity magnification" is a key strength of networks (Provan and Kenis 2008).

Both the social network and resource management literatures discuss how networks influence the capacity of individuals and groups to act. Strong ties between individuals are based on a combination of characteristics, such as intimacy, familiarity, time, and reciprocity (Granovetter 1973). Stakeholders who share strong ties are more likely to influence each other, sharing resources and advice (Crona and Bodin 2006, Newman and Dale 2005). However, strong ties can also create homogenity, as all the actors in the network will know similar things and work in similar ways.

In contrast, new ideas tend to be produced by weak ties. Weak ties tend to exist between dissimilar individuals, offering stakeholders access to diverse pools of information and resources by connecting otherwise separate parts of the network. These ties make a network more adaptable to changes, for example to the political or funding environment, but weak ties, as the name suggests, are easy to break, with the result that individuals sharing weak ties may lack the levels of trust and understanding that are required for collective action (Newman and Dale 2005).

Networks can be managed to improve decision-making or enrich the resources and options available by bringing in different actors and arranging them in specific ways. New actors can be introduced by setting up or re-organizing a network, recruiting them into an existing network, or using them in an advisory role (Kickert *et al.* 1999). The challenge for network managers is to connect actors in ways that enable them to communicate and work together without requiring huge amounts of time or resources. ICT plays a particularly important role in allowing network managers to activate and arrange actors while incurring very low transaction costs. The importance of the ways in which actors are related forms the basis for social network analysis (discussed in Analytics of governance 5.1), which is a tool that can be used to analyze networks and infer their characteristics.

Transnational governance networks

The fact that relations between businesses, governments and NGOs cut across national boundaries is not new (Keohane and Nye 1971), but the importance of non-governmental networks was generally overlooked unless they were directly challenging state authority (Ruggie 2004). Defined as the "regular interaction across national boundaries when at least one actor is a non-state agent or does not operate on behalf of a national government or intergovernmental organization" (Risse-Kappen 1995), transnational governance networks are a key conduit for bringing civil society and businesses into global governance. During the 1990s, a growing number of transnational networks were being organized to act independently of states, leading to their recognition as important agents of change (Andonova *et al.* 2009).

In the environmental sphere, the Jo'burg World Summit on Sustainable Development in 2002 identified partnerships between public, private and civic organizations as the key means through which to achieve sustainable development. In endorsing market mechanisms, the 1997 Kyoto Protocol has stimulated the emergence of networks to support carbon governance. These meetings established networks "as a central steering mechanism" in environmental governance (Pattberg and Stripple 2008: 378).

Within the literature on transnational governance networks, three general types of network have been recognized: epistemic communities, transnational advocacy coalitions, and global civil society networks (Betsill and Bulkeley 2004). An epistemic community constitutes a network of professionals and scientists who adhere to similar scientific and political understandings of a particular topic (Haas 1990), and who work together to influence global political agendas. The network is often maintained by the sharing of factual knowledge and a process of consensual learning. While epistemic communities coalesce around common scientific understandings (epistemology is the study of *how* we know the world), they are often typified by a common political understanding of an issue as well. The IPCC constitutes an epistemic community that shares a scientific consensus around climate change that is used to foster policy change. The epistemic community that addressed CFCs and ozone depletion in the 1980s was also critical in bringing the Montreal Protocol to fruition, as discussed in Chapter 4.

Transnational advocacy networks are comprised of public and private actors, who come together around a specific issue to promote a particular set of actions or viewpoints on it. These networks are bound together primarily by a common set of values, but they also share information and services. Issues that are characterized by polarized positions (i.e. for and against) tend to form the nuclei for transnational advocacy networks (Betsill and Bulkeley 2004). Like epistemic communities, the primary role of these networks is to influence state action, whether at the national or international levels.

By contrast, global civil society networks represent a pure form of governance that takes place beyond the state, comprising groups of non-state actors which create new political spaces. The liberalist school of international relations (discussed in the previous chapter) views these networks as the dominant force within global governance, and sees nation states as mattering only in as far as they facilitate or hamper their formation (Lipschutz 1996).

The Jo'burg World Summit on Sustainable Development in 2002 made a great play of emphasizing so-called "Type II" partnerships between public and private organizations as the best way to implement sustainable development (Glasbergen *et al.* 2007). While this assertion remains largely untested, Type II partnerships are seen as critical in the delivery of sustainable development (Hamilton 2009), and their establishment has gone hand-in-hand with the emergence of

Analytics of governance 5.1

Social network analysis

Like ANT, discussed in Chapter 3, social network analysis is based on a relational ontology. Rather than focus on the status of actors in a network, or the nature of the relation between actors, social network analysis simply represents the presence or absence of a tie and the relative strength of that tie. Data are typically generated through structured interviews, questionnaires, or observation of network participants, which interrogate specific types of relation, for example, based on information exchange, authority or trust. Recording information about the number and strength of ties in quantitative form makes it easy to represent the results graphically, which can then be used to produce visual representations of social networks (UCINet and Netminer are among the most commonly used pieces of software in academia).

Social network analysis reveals the levels of connectivity and centrality in a network. Figure 5.1 shows four simple networks with different levels of connectivity, measured by levels of reachability (the degree to which all nodes are connected), and density (the number of connections each node has). The networks in the top half of the figure have high levels of

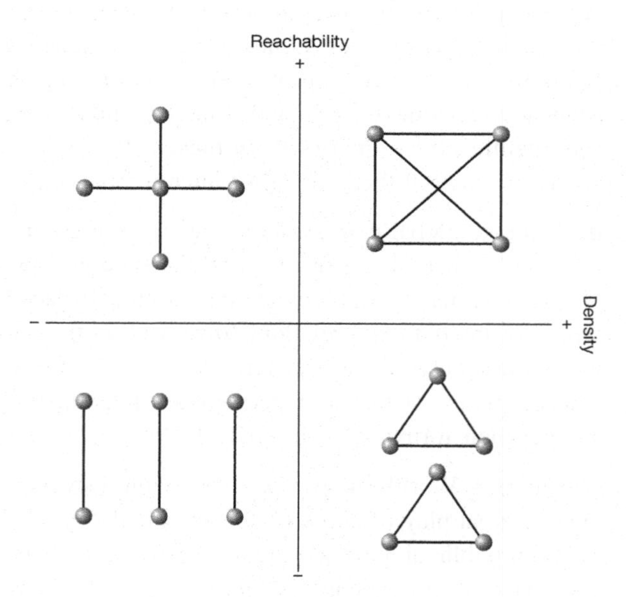

Figure 5.1 ***Social networks with high and low levels of connectivity, as indicated by reachability and density***

Source: Janssen *et al.* 2006.

connectivity, which means that information and innovations can be diffused quickly, and stakeholders can align their interests and working methods more accurately. The disadvantages of being highly connected are that bad practices or pathogens can spread very quickly, making the network brittle. Being highly connected produces a version of the embeddedness paradox, whereby actors need to be embedded in order to work effectively together, but are less likely to innovate by sheer dint of being embedded (Uzzi 1997). The networks in the lower half of the figure have low levels of connectivity, which gives them the potential to form dense clusters that respond to problems in distinctive and complex ways. Clustering breeds innovation and resilience to changes in political or economic conditions, but makes it hard to access and spread information across the network.

By contrast, Figure 5.2 shows social networks with high and low levels of centrality. Highly centralized networks make it easier to coordinate collective action, because there is a central actor connected to all others. High levels of centrality also have the potential to make the network more accountable, as the central actor can be held responsible for the actions of the network as a whole. The disadvantage of high levels of centrality is that the system is more vulnerable if the central actor leaves or is weakened. Further, highly centralized networks are more rigid and hierarchical, making them appear less democratic and fair (Janssen *et al.* 2006). Networks with low levels of centrality can be more inclusive of different groups, and are highly resilient to the loss of specific actors, but lack accountability and can be inefficient at solving simple problems due to the lack of overall coordination.

By revealing the structure of networks, social network analysis identifies which stakeholders are more important, which are marginal, and how stakeholders cluster together. Ties can also be visualized, including whether they are reciprocal (two-way) and how strong they are (the thickness of the line). By quantifying the extent to which the stakeholders trust one another, social network analysis can identify issues between them, providing a basis for management interventions to enhance information flows where necessary, or to select stakeholders to work together (Prell *et al.* 2009).

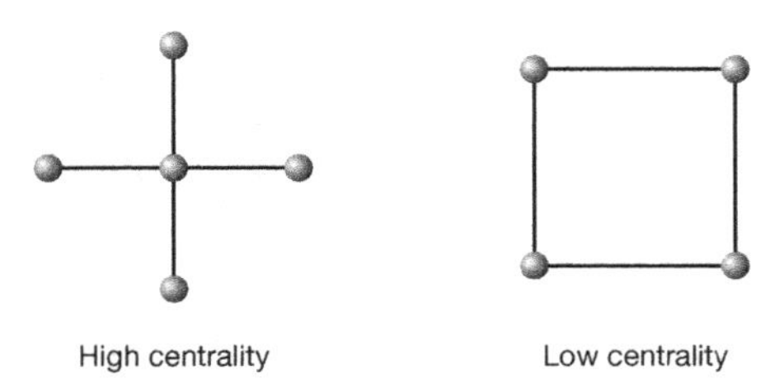

Figure 5.2 ***Social networks with high and low levels of centrality***

Source: Janssen *et al.* 2006.

transnational networks (Andonova and Levy 2003). Such networks can involve public bodies, private bodies, or a combination of both, and Table 5.1 gives examples of public, hybrid and private transnational networks that have emerged around the issue of climate governance. Purely public networks involve only state actors, like the C40 network that brings 40 large cities together in order to enable them to exert a greater influence over climate governance on the global stage. The Type II partnerships promoted at the Jo'burg World Summit on Sustainable Development are hybrid networks that bring public and private bodies together to address environmental goals. Private networks usually involve some form of business self-regulation, often coordinated by NGOs and funded by governments. The Renewable Energy Policy Network for the 21st Century, and Carbon Disclosure Project are all discussed as case studies later in this chapter.

While distinctions between public, hybrid and private networks can be hard to draw in practice (most lean one way or the other, but few could be considered pure examples), all link organizations together to do things they otherwise could not or would not be able to do. Transnational governance networks thus "form an increasingly dense layer of governance, which can be compared to a transmission belt, linking governance systems from the global to the local, as well as across the public and private spheres" (Andanova *et al.* 2009). The Renewable Energy Policy Network for the 21st Century, discussed in Case study 5.1, is an example of a transnational network that builds capacity to achieve considerable results with limited resources.

Table 5.1 ***Transnational networks of climate governance***

Public	*Hybrid*	*Private*
Governmental (e.g. C40 Cities for Climate Protection Campaign)	Type II partnerships (e.g. Renewable Energy Policy Network for the 21st Century)	Businesses and NGOs (e.g. Carbon Disclosure Project)

Source: adapted from Pattberg and Stripple 2008 and Börzel and Thomas 2005.

Case study 5.1

Renewable Energy Policy Network for the 21st Century

The Renewable Energy Policy Network for the 21st Century (REN21) is a global network that supports the adoption of renewable energy through policy work, advocacy and information exchange. Originating in the Political Declaration of the International Conference for Renewable Energies in Bonn, Germany, 2004, REN21 aims "to work within a 'global policy network' with representatives from parliaments, local and regional authorities, academia, the private sector, international institutions, international industry associations, consumers, civil society, women's groups, and relevant partnerships worldwide" (REN21 2010). Its official origins and direct funding from the German government lent REN21 early legitimacy and helped it to grow quickly, enrolling a vast range of stakeholders in to its network.

REN21 employs fewer than 10 people who run its entire network on a budget of only US$1 million per year. These efficiencies are a product of its internal organizational structure, which is designed to promote capacity magnification. Overall strategy is set by the steering committee, which comprises a broad coalition of influential and informed people who are active in the international renewable energy arena. Their work is supported by a permanent bureau consisting of members from the steering committee and the secretariat, which is charged with taking interim decisions. Having such a broad coalition of influential members on the steering committee is vital for the success of the network, placing it at the cutting edge of policy developments and significantly magnifying its capacity to influence policy-making. REN21 uses its members to promote its agenda at UNFCCC CoP meetings, host high-profile international events, and produce influential issue papers (notably, its *Renewable Energy Global Status Report*). The network also hosts an open forum for information exchange and discussion on its website.

The REN21 network is characterized by weak ties, with little formal control over its members, and has no official rules that must be adhered to (Bugler *et al.* 2010). Any institution, organization, government, or even anyone with access to a computer, can join. Members are guided by the agenda set by the steering committee, and the circulation of information is used to generate a community among its members. While the network is open in terms of membership, allowing a wide range of views to be expressed and a large volume of information to flow around the renewable energy community, some issues of accountability and legitimacy present themselves when looking at the network structure. Table 5.2 lists some of the strengths and weaknesses of REN21, which relate to the wider characteristics of network governance, discussed at the end of the chapter. The strengths revolve around the ability of the network to influence high-level policy processes with relatively few resources, while the weaknesses concern the transparency with which this is done. For example, the network is governed by a largely unelected and unaccountable steering committee, and lacks clear boundary rules concerning how actors can gain access to these positions.

Table 5.2 ***Strengths and weaknesses of REN21***

Strengths	*Weaknesses*
Steering committee employed by other organizations, allowing the network to save huge sums of money	Strategy largely dictated by steering committee, but lack of transparency as to how members are chosen
Deliberate spread of representatives in terms of expertise and geographical area ensures global scope	Submissions to the website approved by the secretariat, which comprises a group of 10 unelected employees
Connections with other networks used to cooperate on specific projects	Funding comes from the German government, questioning the independence of the network
Connections of members used to get REN21's policy priorities discussed at major conferences	The network is not a legal entity—it is unclear who is ultimately responsible for actions undertaken on its behalf

Corporate social responsibility

While it is unquestionable that private industry has a major role to play in addressing environmental problems, the question of how to alter its current activities is less clear. Environmentalists tend to favor tighter regulations, but the model of network governance privileges cooperation and voluntary action. While government regulatory strategies are normally presented as driving changes in industry, whereby compliance with legislation provides the baseline for environmental performance, corporate self-regulation is in vogue, and increasingly businesses are engaging voluntarily in environmental governance networks that help them certify practices or products as sustainable, or reduce their environmental impact. Advocates suggest that self-regulation is more effective as businesses are in a better position to determine how to effectively control their actions than a government regulator. Self-regulation also has a number of potential benefits for business: delaying or weakening new legislation; increasing the legitimacy of the business; and acting as a focus for best practice. At the same time though, self-regulation is voluntary and thus open to abuse.

Notions of environmentally friendly business have historic precedents in firms such as Cadbury's (recently taken over by Kraft), which believed they had a duty to improve society. As the most recent Cadbury (2002) says,

> the continuing existence of companies is based on an implied agreement between business and society . . . The essence of the contract between society and business is that companies shall not pursue their immediate profit objectives at the expense of the longer term objectives of the community.

Corporate social responsibility (CSR) has emerged as a voluntary commitment by businesses to ensure that their operations do not run counter to the wider good of society and the environment (Blowfield and Murray 2008).

The World Bank (2004) describes CSR as

> the commitment of business to contribute to sustainable economic development, working with employees, their families, the local community and society at large to improve their quality of life, in ways that are both good for business and good for international development.

CSR is based upon a stakeholder model of the firm, whereby businesses are seen as groupings of shareholders, customers, workers, the community of which they are a part, and so on. Rather than impose legal requirements to mitigate the social and environmental impacts of business practices, CSR enables companies to govern themselves.

A number of CSR indices have been developed in response to demand from ethical investors. For example, the FTSE4Good and Dow Jones Sustainability Indices judge companies on criteria like human rights, stakeholder relations and their environmental impact. The UN has an initiative called the Global Compact, which is a voluntary international corporate network to support the participation of both the private sector and other social stakeholders to "advance responsible corporate citizenship and universal social and environmental principles to meet the challenges of globalization" (United Nations 2004). The Global Compact has 10 principles organized around human rights, labor standards, environment, and anti-corruption. The environment principles state that businesses should support a precautionary approach, undertake initiatives to promote greater environmental responsibility, and encourage the development and diffusion of environmentally friendly technologies.

The most common criticism of CSR is that companies will engage in order to improve their image, without stopping profitable but environmentally damaging activities (Vogel 2006). So-called "greenwash" is undoubtedly an issue, as companies become involved with various environmental initiatives purely to generate good publicity (Moneva and Archel 2006). Even before the disastrous Gulf of Mexico spill in 2010, Greenpeace gave their Emerald Paintbrush Award for Greenwashing to BP in 2008 for their rebranding exercise from "British Petroleum" to "Beyond Petroleum." Although the company's carbon emissions were reduced by 10 percent, its $20 million investment in sustainability measures yielded $650 million in savings and increased sales of natural gas. Further, it swiftly discontinued its carbon trading scheme when this turned out to be unprofitable, and continues to pour 93 percent of its investment into oil and gas exploration, compared to only 1.3 percent for solar energy.

More radical scholars suggest that businesses are simply afforded too much latitude by CSR, and highlight how regulations to ensure that corporations served the public interest were progressively removed in the nineteenth century, leaving them today with greater legal rights than people and yet none of the responsibilities associated with being a citizen. Writing in the mid-twentieth century, Karl Polanyi (1944) argued that the separation of the economy from society, facilitated primarily by the idea that markets should be free from regulation, was a mistake. By definition markets lack a social conscience and are thus incapable of self-regulation.

Market advocates don't like CSR either, arguing that it is "a dangerous distortion of business principles" (Drucker 2004). Corporate behavior should be motivated by the pure pursuit of profit, within the constraints of the law, in order to allow market forces to operate efficiently. There is no doubt that corporations may not be the best equipped organizations to deliver wider social and environmental benefits. On a more fundamental level, CSR is limited by the lack of clear political and legal framework coordinating the very thing it is supposed to be helping—society. Without a clear notion of what rights and responsibilities exist, it is hard to develop a clear picture of what an effective CSR policy might be and how it might be monitored (Ramus and Montiel 2005).

Certification networks

Perhaps the most important way in which companies are being enrolled into environmental governance is through certification networks. Reporting and accreditation is a central mechanism through which corporate self-regulation takes place in general, and was seized upon by environmental NGOs and governments after the Rio Earth Summit in 1992 as a way to bring economic, social, and environmental issues together. Having a certified production process provides an organization with a quality stamp of approval. The oldest eco-labeling scheme is Germany's *der Blaue Engel* (the Blue Angel), which has been awarded since 1978 to companies who make significant commitments to environmentally friendly practices in both production and consumption. *Der Blaue Engel* had certified 4,000 products by the early 1990s, and the Ecolabel Index, in conjunction with the World Resources Institute, currently tracks 349 eco-labeling schemes, covering 212 countries and 37 industrial sectors. The phenomenal growth of these schemes reflects the existence of a market for products that are certified, as people are motivated to change their consumption behavior through innovative communication strategies and better branding of sustainable lifestyles.

Some of the most interesting and influential transnational certification networks encourage sustainable practices among private business (Gulbrandsen 2010). The Forest Stewardship Council, discussed in Case study 5.2, has been exceptionally successful in using voluntary certification to improve the sustainability of the forestry industry.

Auditing networks

There is an increasing feeling among policy-makers that in order to make sustainability happen it must be measured. Great efforts have been expended developing ways to audit corporate practices against specific, measurable sustainability criteria (Bennett *et al.* 1999). Environmental auditing is a decision-making tool used primarily in business and industry that focuses on the sources of environmental impacts, rather than the effects, involving a systematic examination of environmental information about an organization, a facility or a site to verify whether it conforms to specified audit criteria. The process emphasizes continual improvement rather than measuring environmental impact relative to an independently set standard or threshold (Petts 1999).

Environmental and sustainability auditing is dominated by the EU's Eco-Management and Audit Scheme and the UN's Global Reporting Initiative. The Eco-Management and Audit Scheme is a voluntary initiative designed to improve the environmental performance of companies. It aims to recognize and reward those organizations that go beyond minimum legal compliance and requires participating

Case study 5.2

The Forest Stewardship Council

More than any other organization, the Forest Stewardship Council (FSC) has made chopping down trees environmentally friendly. Established in 1993, it has certified some 134 million hectares of commercial forest in over 80 countries as sustainable, and helped to ensure that entire supply chains from tree to customer are managed sustainably. Its distinctive tick-tree logo will probably be on the next wooden product that you buy. Surprisingly, all this has been achieved without any legal regulations in less than 20 years. The FSC is a great example of the power of network governance to effect change—companies have signed up to the certification scheme voluntarily —and shows how transnationalization can actually lead to local activities facing more scrutiny over their environmental activities, not less.

Conceived in 1990 and formed after the Rio Earth Summit in 1993, the FSC was the brainchild of a group of timber users, traders and environmental and social NGOs who were interested in setting up a system to certify timber products that were sourced from sustainably managed forests (Eden 2009). The group originally lobbied key countries involved in the Rio Earth Summit to adopt a certification scheme, but when the conference failed to reach a binding agreement on deforestation the FSC decided to press ahead with its plans, securing funding from the World Wildlife Fund and DIY giant B&Q to set up an office of three people in Oaxaca, Mexico, in 1994. By 2003 the FSC had grown to 25 staff, moved to Germany, and established its tick-tree certification logo as a familiar sight in stores around the world. The FSC is now funded by a range of organizations, including other charities, governments, companies with an interest in home improvements like IKEA and Home Depot, membership subscriptions, and fees from certification bodies.

The FSC represents an interesting case of private governance, which is non-state and market-driven (Cashore 2002). In other words, it brings together the interests of environmentalists and business, and exercises authority in regulating and enforcing its own policies and environmental standards in the absence of any direct state involvement. Authority is established by the approval of external audiences, like the state, environmental NGOs and, most importantly, consumers through their market choices. Cashore (ibid.) suggests that the legitimacy of such networks is pragmatic, in that the network delivers substantive benefits to its members, moral, in that it is "the right thing to do," and cognitive, in that to do otherwise is literally "unthinkable." That said, research highlights that members are motivated primarily by pragmatic rather than moral considerations, and that perceived legitimacy does not necessarily mean that a network is contributing to sustainability (Bernstein and Cashore 2004).

organizations to produce a public environmental statement that reports regularly on their environmental performance. The Global Reporting Initiative is a UN scheme which brings representatives from business, accountancy, investment, environmental, human rights, and research and labor organizations from around the world together to develop and disseminate globally applicable sustainability reporting guidelines. These guidelines are voluntary, enabling organizations to report on the economic, environmental, and social dimensions of their activities, products, and services. Started in 1997, the Global Reporting Initiative became an independent organization in 2002, and is an official collaborating center of the United Nations Environment Programme, working in cooperation with former UN secretary-general Kofi Annan's Global Compact.

Levels of reporting vary between country and industrial sector because the level of public pressure varies (for example, companies in some sectors or countries have more environmental pressure groups looking over their shoulder) and the policy context varies (for example, some sectors or countries have stricter legislation so companies are less keen to try to outperform this high minimum level).

The Eco-Management and Audit Scheme and the Global Reporting Initiative are far from the only environmental reporting networks in existence. As climate change grows in importance, numerous networks are being established to encourage organizations to engage in carbon reduction activities. An example of a smaller, more dynamic initiative is the Carbon Disclosure Project (discussed in Case study 5.3), a not-for-profit NGO network which measures and discloses corporate climate change commitments.

It is hard to evaluate the overall success of certification and auditing schemes in greening business. Private companies are a highly heterogeneous group, varying in terms of their activities, size, and level of environmental concern and action, and although some firms really have made a difference, others are certainly guilty of greenwash. While the success of certification schemes like the Forest Stewardship Council brings the role of private industry as the villain of environmental change into question, a UNEP (2004) survey found that 50,000 multinationals still fail to report on the environmental impact of their activities, and that most of those who do report take no action or fail to link it with financial reports. That said, the phenomenal growth of certification and auditing networks represents one of the most dynamic trends in environmental governance, and there seems to be an almost exponential demand for disclosure, to the extent that the market is almost saturated with competing agencies counting on businesses to volunteer their information (Park *et al.* 2008).

Case study 5.3

The Carbon Disclosure Project

The Carbon Disclosure Project was launched in 2000 in London, and has grown ten-fold from 235 responding companies in 2004 to more than 3,000 in 60 countries in 2010. It collects information on greenhouse gas emissions, water management and climate change strategies from its members, and makes this information available to more than 534 financial investors holding some $71 trillion of assets, in order to help them make more sustainable investment decisions. The Carbon Disclosure Project also works closely with governments to improve the sustainability of public procurement, and has recently begun collecting data from cities.

As with many environmental networks, the Carbon Disclosure Project works through partnerships to magnify its capacity. Major IT companies like Accenture and Microsoft have helped build its online database, while the financial information giant Bloomberg has incorporated Carbon Disclosure Project data into its live feeds. Its operations in other countries are coordinated by partners who are not directly employed by the Carbon Disclosure Project, and the organization has multiple income streams. The figures from its own website list 30 percent of funding as coming from corporate sponsorship, 28.7 percent from special projects, 18.3 percent from grants and donations, 15.1 percent from international partnerships, 5.1 percent from memberships and 2.8 percent from other sources.

The Carbon Disclosure Project is independent insofar as it is beholden to no single authority, but has to take the priorities and preferences of its key funders into account. By necessity, the network must resonate with the wider objectives of climate governance, but, equally, it must also provide a positive spin for the activities of its membership or face ruin. Such is the leitmotif of network governance—connected but compromised. Additionally, the information that the Carbon Disclosure Project collects is not independently verified but measured by the member organizations themselves, raising questions about its accuracy. But, setting these practical drawbacks aside, the phenomenal growth of the Carbon Disclosure Project indicates the power of disclosure to generate change, and the appetite of the corporate sector to engage with voluntary networks.

Conclusions

Table 5.3 summarizes the strengths and weaknesses associated with the network mode of governance. Perhaps most obviously, voluntary networks do not have the political authority of a traditional nation state. As Cashore (2002) notes in relation to non-state, market-driven networks like the Forest Stewardship Council, there are no democratic elections and no one can be fined or imprisoned for failing to obey the rules. While networks allow actors to pool resources, enabling them to do things they would not otherwise be able to, motivation to join is based purely on self-interest and there are few formal constraints preventing actors from leaving the network, making them less robust.

The growing influence of NGOs and companies in the environmental sphere raises a series of questions concerning their accountability and representativeness in decision-making. Quasi-governmental organizations and NGOs exert considerable power within governance networks without being either directly elected or directly accountable to the public (Weber and Christopherson 2002). Further, network governance may actively undermine elected governments, jeopardizing political equality and individual liberty as conflict occurs behind closed doors rather than in the public spaces of political debating chambers

Table 5.3 ***Strengths and weaknesses of network governance***

Strengths	*Weaknesses*
Collective and reflexive	No real political power
Widens representation	Ineffective as decision taken in advance (becomes a public relations exercise)
Broadens participation	Non-accountability of non-state actors and capture by dominant interests
Consensus (conflict resolution)	Compartmentalization of policy
Innovative restructuring of institutions	Dominance of expert and industry knowledge due to complexity of problems
Recognizes complexity of real world	Disperses responsibility for making change happen
Diversity of institutions	Turf wars over areas of operation

(John and Cole 2000). As Cornwall (2004) notes, networks are often not popular spaces where anyone can join in, but "invited" spaces with carefully policed boundary rules. Further, there is a danger that far from producing innovative answers, networks simply reproduce dominant ideas, as they strive to keep donor organizations happy (Taylor 2007). From a governance point of view, this can allow governments to use networks to carry out their own pre-determined agendas (Klijn and Skelcher 2007). While networks are critical in implementing environmental agreements, there is a danger of saturation in certain areas, as multiple institutions seek to do very similar things. As Bulkeley and Newell (2010) argue, this can lead to confusion and conflict between networks.

On the other hand, network governance responds to the democratic deficit in traditional parliamentary systems, affording a larger proportion of stakeholders a voice in more decision-making processes (Sorensen and Torfing 2007). In principle, anyone can set up an instition or network, as the proliferation of networks around an issue like climate change attests. Networks are increasing the layers and clusters of non-state rule-making and rule-implementation, both vertically and horizontally, that run alongside the traditional system of legal treaties negotiated by states (Tienhaara 2009).

Reviewing 137 cases of collaborative governance across a range of policy sectors, Ansell and Gash (2008) identify five critical factors that influence the success of network governance: prior history of conflict or cooperation, incentives for stakeholders to participate, imbalances of power and resources, leadership, and institutional design. Collaboration itself requires face-to-face dialogue, trust building, and the development of commitment and shared understanding. The authors found that a virtuous cycle of collaboration tends to develop when networks focus on "small wins" that deepen trust, commitment, and shared understanding. Obviously these conditions either do not or cannot exist in all cases, but their final factor, institutional design, is critical in addressing the prior four through setting appropriate rules, selecting the right stakeholders and actively managing networks.

Scholars have also studied whether seemingly functional networks achieve the kinds of things that their exponents argue. For example, Betsill and Bulkeley's (2004) study of the Cities for Climate Protection program questions the received wisdom that transnational networks primarily facilitate the exchange of knowledge and information. Instead, they found that local governments were mobilized more by the financial resources on offer, and the political legitimacy conferred by being visibly involved with climate protection, than by access to information. Similarly, placing our faith in the ability of networks to coerce business into acting voluntarily ultimately depends to some degree on the preferences of consumers. The development of 4×4 sports utility vehicles in the 1990s was a crass failure of the automobile industry to voluntarily address climate change, but it was driven by consumer demand. Similarly, eco-labeling schemes depend on consumers caring enough to potentially pay more for certified products. Even the voluntary aspect of CSR is often less than it seems, with shareholders, customers and investment funds demanding evidence of environmental sustainability.

Network governance increasingly blurs the distinctions between the state, NGOs, private companies and the public. For example, governments must increasingly participate in networks in order to find out about and exercise the most up-to-date and effective forms of regulation. Conversely, Bäckstrand (2008) has pointed out that the perceived shift from a "sovereign to post-sovereign" world, where states are becoming less influential, is at odds with much of the governance literature. Transnational networks are said to operate in "the shadow of hierarchy," whereby states remain influential because they have the power to delegate rule-setting functions to partnerships and networks. Making a similar point about business, Berry and Rondinelli's (1998) study of the way in which companies voluntarily implement pollution control technologies highlighted the influence of the state, through increasing legal liability and the cost of waste disposal.

The next chapter looks at the market approach to environmental governance, which, rather than depending upon actors to engage in collective action voluntarily, seeks to motivate actors on the basis of financial loss and gain.

Questions

- What is the relationship between transnational environmental governance networks and the state?
- Do networks need to be accountable in order to address environmental problems?

Key reading

- Banerjee, S. (2008) "CSR: the good, the bad and the ugly," *Critical Sociology*, 34: 51–79.
- Bulkeley, H. and Betsill, M. (2004) "Transnational networks and global environmental governance: the Cities for Climate Protection program," *International Studies Quarterly*, 48: 471–93.
- Klijn, E. and Skelcher, C. (2007) "Democracy and governance networks: compatible or not?" *Public Administration*, 85: 587–608

Links

- https://www.cdproject.net/en-US/Pages/HomePage.aspx. Home of the Carbon Disclosure Project.
- http://craneandmatten.blogspot.com/. A thoughtful if critical blog on corporate social responsibility.

6 Markets

Intended learning outcomes

At the end of this chapter you will be able to:

- **Understand the basic principles of the market approach and how markets are used to address environmental problems.**
- **Evaluate the market mechanisms associated with the Kyoto Protocol.**
- **Appreciate the power of placing financial values on the environment.**
- **Be conversant with the strengths and weaknesses of market approaches to environmental governance.**

Introduction

> We can't solve problems by using the same kind of thinking we used when we created them.
>
> (Albert Einstein, 1879–1955)

The greenhouse gas emissions that are currently causing climate change have been produced primarily by industrial activity associated with the massive expansion of the global economy over the last 250 years. Rather uncomfortably for those seeking to argue that economic development can be sustainable, industrial output correlates almost perfectly with greenhouse gas emissions, so that more economic activity equals more emissions. One startling indication of this relationship is that the 2008 global economic recession did more to reduce emissions than the efforts of environmentalists and regulators put together. On this reading, it is questionable whether the brand of market economics that caused climate change is capable of reversing it.

But the flipside of this analysis is that markets represent the most important lever we have to reduce the impact of society on the environment. If markets can be designed in the right way, then environmentally friendly behavior can be aligned with the most profitable actions for companies and consumers. In contrast to other modes of governance, which coerce groups of actors to take voluntary measures, markets coordinate individual actions through the manipulation of the profit motive. The abortive attempts of the international community to forge a binding global agreement on greenhouse gas emissions reductions stand in stark contrast to the enthusiasm among all major emitting nations for emissions trading markets. But can markets be transformed from an ecological scourge into an environmental savior?

This chapter considers how environmental goods such as clean air and water are increasingly being incorporated within markets. Previously, many common environmental resources have simply been used for free, leading to what economists call "negative externalities"—unintended economic impacts that are not included within the costs of production. Climate change can be seen as the negative externality *extraordinaire*, generating huge costs associated with freak weather events, sea level rise and so on that were never included in the original cost of fossil fuels. Market approaches seek to include the costs of negative environmental externalities within prices, arguing that if the costs of using common environmental resources can be valued then they will be protected. The chapter then considers how this logic is put into practice, including the ways in which common environmental resources can be captured in market valuations. Particular attention is paid to markets for carbon emissions, as they present the most ambitious attempt to apply market principles. The chapter considers the power of placing financial values on the environment as a tool to aid its governance, and concludes with a consideration of the strengths and weaknesses of the market approach.

Using markets

Markets solve the tragedy of the commons by turning them into private property. Privatization divides common resources into discrete packages of property, which are then allocated or sold to individuals and groups. In the absence of a strong collective urge to protect common resources, private ownership provides the motivating force as people seek to protect what is theirs (Stroup 2003). Put simply, no commons equals no problem. The same logic holds for what economists call "negative externalities"—the harmful side-effects of activities that are not taken into account beforehand. Markets can be designed to include the cost of the atmosphere, which was formerly freely available to pollute, in the overall costs of production. So, climate change can be addressed by making individual polluters purchase the appropriate amount of atmospheric capacity to absorb their greenhouse gas emissions. The logic is that making industry pay for the full cost of its activities will prompt it to adopt less polluting technologies (for example, replacing coal-fired power stations with wind farms).

Some commentators are more pessimistic, arguing that externalities are pervasive in the market approach. To take the example of agriculture, the logic of competition dictates that larger producers are more successful due to the economies of scale that they can achieve. Over time, diverse local producers are progressively rationalized into larger specialized operators who can produce more units at a cheaper cost. At the extreme end of this process one finds a series of undesirable outcomes, like the farm in Utah that houses one and a half million pigs and produces more sewage than the city of Los Angeles. This creates a massive extra economic cost of dealing with the sewage problem, and generates huge energy and water demands (the pigs are all kept inside). Beyond the economic problems of concentration and specialization there is a range of wider social problems: the impacts on standard of living for nearby residents who have to live with an overpowering stench; the problem of animal welfare in factory farming systems; unpleasant and dangerous conditions that must be endured by the workforce; lower quality food products, and so forth (McKibben 2007).

Market advocates do not deny the existence of such problems, but argue that traditional regulation produces its own *political* externalities, whereby too many resources are preserved. The process of using markets in the environmental field is one of trial and error to get the balance right—as one set of advocates says, "mistakes will be made" (Anderson and Leal 2001: 22). That said it is important to note that markets are better at capturing some externalities than others. Drawing on work by Farber (2007), Neil Adger (2010) notes that markets tend to work better for geographically constrained impacts, where externalities are confined to a body of water, coastline, or habitat, but less well for diffuse impacts (for example, on global food systems) or for catastrophic climate changes at the global level. Markets are also not good at including social externalities, like the adverse impacts on communities or places, and losses of non-material assets, like the beauty of a landscape that may be destroyed by resource extraction.

Markets assume that private actors (individuals or organizations) constitute the basic units of society, and that they behave rationally to maximize their own benefit in accordance with the best information that they have. More extreme market advocates, like neoliberal economists, argue that the role of the state is simply to allow individuals to be able to act in their best interests by freeing markets to take their own course. The eighteenth-century Scottish economist Adam Smith famously referred to these self-guiding qualities of the market in *The Wealth of Nations* (1776) as the "invisible hand," whereby individuals pursuing their own gain will be "led by an invisible hand" to promote the public interest. For market advocates, the role of the state is simply to ensure that the legal barriers to establishing markets are minimal, and that private individuals are allowed free rein to trade environmental goods in the market place, in order to maximize the "efficiency" of the market. As Anderson and Leal (1991: 4) state, "instead of intentions, good resource stewardship depends on how well social institutions harness self-interest through individual incentives," essentially by creating markets in which the most profitable behaviors are aligned with those that deliver desirable environmental outcomes. Good market design steers collective action by incentivizing private actors to undertake certain activities.

The idea that individuals are rational economic actors is closely related to the efficient market hypothesis, which holds that markets are the best way to reach decisions because they pool knowledge in the most effective way. If an outcome is uncertain, as it often is in the environmental sphere, then multiple knowledges will exist about a situation, making centralized decision-making inefficient. Even something as simple as next season's coffee harvest cannot be predicted accurately due to the vagaries of climate and Latin American politics. In this case, a system of market exchange allows actors with different knowledges and concerns to interact seamlessly, producing collective decisions through the setting of prices according to supply and demand. Markets thus provide multiple, fast feedbacks in the form of prices. In governance terms, collective action is coordinated by the rules of market exchange, rather than by regulatory control (as it would have been in the era of command-and-control policy) or common understanding (as it would be in a voluntary network).

Markets pool information concerning the way in which environments are valued in a similar way. This is important, as many environmental management questions depend on what we value. As Hardin (1968) asked in relation to the tragedy of the commons, "We want the maximum good per person; but what is good?" For example, forests do not in themselves dictate how they should be managed—timber production, recreational activities, wildlife habitat, and aesthetic quality are all legitimate uses that must be balanced against one another (Anderson and Leal 1991). Ecological science and mathematical efficiency models can help maximize benefits, but the question of which benefits should be maximized depends upon human preferences. Markets reveal the preferences of individual humans through the prices that they will pay for different things.

Further, economists argue that because markets transmit future concerns over scarcity into current prices, they drive innovation (Solow 1974). Substitutability is the idea that as a resource becomes increasingly scarce its price will rise, forcing alternatives, or substitutes, to be found. This is a fundamental premise for those who advocate adaptation to climate change in the future rather than mitigating against it now. Substitutability assumes that markets and technology are ingenious enough to replace the resources that we use up, for example, replacing fossil fuels with renewable energy, developing genetically modified organisms to replace plants that can no longer survive on a warmer planet, using single cell worm protein instead of animal and fish protein, and spreading iron filings in the sea to replace the forests that used to sequester carbon. Nobel Prize-winning economist Solow once stated this thesis in its purest form: "[i]f it is very easy to substitute other factors for natural resources, then there is, in principle, no 'problem'. The world can, in effect, get along without natural resources" (1974: 11, quoted in Walker 2009). While far from unproblematic in technical terms, the real question here is ethical. Do we want to live on a planet where nature has died?

While it is not hard to find faults with the efficient market hypothesis (the 2008 financial crisis), or rational man (consider the way commercial advertising plays on our emotions), proponents of markets tend to see them, if not as perfect, then as the best approach available to us. The failure of Soviet-style central planning in the twentieth century, including the environmental devastation that accompanied it, is often held up as evidence that models of society which do not have efficient feedbacks between supply and demand tend to fail (Perrings 1998).

Economists see market assumptions as ideals that best mirror human behavior and thus deliver the most desirable social outcomes.

Enclosure and commodification

For market advocates, the role of institutions is simply to create markets in environmental goods that allow them to be traded like any other good. To privatize a common resource like land or water, the resource must be enclosed into privately owned parcels. Sometimes enclosure has been quite literal. For example the communities who settled the American West were able to create individual farmsteads with the invention of cheap and durable barbed wire in the 1870s, which allowed them to partition off vast tracts of land. In other cases, enclosure is more abstract. For example, the creation of markets for extraction from aquifers grants private individuals rights over a specified amount of water, rather than ownership of a specific set of water molecules (Cowan 1998). Efforts to allocate fishing quotas in the EU represent a classic case of the difficulties of enclosing environmental goods. Fish simply do not respect national boundaries, while the system of maritime sovereignty is highly complex (Bear and Eden 2008). Entire communities depend on fishing to survive, and the question of who actually "owns" specific fish has caused a number of stand-offs between national fishing fleets around the world.

Ensuring that common resources are split up fairly is a major governance challenge. The Enclosures Acts of the eighteenth and nineteenth centuries in Britain, which transformed common agricultural land into the archetypal English landscapes painted by Turner and Constable, was often a violent process whereby the aristocracy simply evicted peasants and seized common land. Where common resources are subject to multiple claims the process of granting rights in a way that is acceptable to all parties can be nigh on impossible. As discussed in Chapter 4, the main stumbling block to reaching a global agreement on reducing greenhouse gas emissions concerns whether developing countries should be bound to reduce emissions as well as developed countries. In market terms, this boils down to how to allocate rights over the atmosphere.

Precedent use, which takes into account existing dependence upon a resource, is often used to determine the need of different parties. So, for example, the EU emissions trading scheme allocated carbon credits to existing polluters on the basis of how much carbon they were already using. Companies that were creating most pollution (i.e. "using" the most atmosphere) received the lion's share of the resource. While this respects continuity with the past and minimizes disruption to existing activities, it also runs the risk of perpetuating undesirable activities and long-standing injustices. So, for example, companies that have already taken steps to reduce their emissions are effectively punished, as they will receive fewer credits, whereas companies who have made no effort to lower their emissions will benefit. Returning to the question of greenhouse gases, the developing world argues that the USA should do most to reduce emissions, as it has already "used" more than its fair share of the atmosphere, whereas the USA argues that precedent use should be taken into account and thus it should be granted a higher per capita emissions allowance.

Enclosure privatizes a resource, but in order to trade it in a market the units that are created must be fungible, i.e. interchangeable and equivalent to one another. This is problematic in the environmental sphere, as ecological processes are often linked to the places in which they occur. For example, in the late 1990s the USA experimented with a system of wetland banking, whereby developers could destroy wetlands if they purchased a similar area of wetlands that were created elsewhere (Robertson 2004). Wetlands are highly specific in terms of their ecological function, though, making it hard to establish equivalence between two geographically distant sites. Location matters—a wetland next to a human settlement will have higher recreation utility as more people will be able to visit it, and its ability to soak up rainfall and reduce flooding will also be more valuable because it will protect more property. As Bakker (2005) notes in relation to water, commodification is not the same as privatization—it is so fluid that it resists the logic of exchange. Creating units for exchange does not mean that they can be exchanged. While economic valuation concerns statistical units, ecosystems are embedded in specific places, making the task of creating fungible units complex and expensive.

Despite the difficulties of enclosing something as fluid as air, the atmosphere is becoming an increasingly commodified and privatized resource. Thornes and Randalls (2007: 2, after Castree 2003) identify what they call a "new atmospheric paradigm" in which atmospheric services are being financialized, characterized by instruments like weather derivative trading that allow organizations to insure against losses due to inclement weather. Traders can use offset derivatives to make a profit independently of what the weather actually does. So, for example, an ice cream seller may insure against a cold summer, which will hit sales, whereas a building operator may insure against a heatwave, which will raise air conditioning costs. The broker can charge both and balance the losses and gains of each against the other. The Weather Risk Management Association (2010) estimated the value of weather derivatives traded in the year 2005–6 at $45 billion, which compares to a total global spend on climate and meteorological research of around $10 billion in 2002.

Advocates also argue that markets can dictate when it becomes necessary to establish property rights within a system of resource use. If the economic costs of depleting a common resource outweigh the economic costs of setting up and regulating a market for that resource, markets will simply appear as the resource will have become scarce enough to have value (Anderson and Leal 2001). The counter-argument, of course, is that a global resource like the atmosphere may already be irreversibly damaged by the time its worst effects become felt.

Using markets to regulate environmentally damaging behavior is a more complex process than simply bartering fruit on a street stall. These complexities are explored further by looking at the market approaches associated with the Kyoto Protocol.

Evaluating markets: Kyoto and beyond

The Kyoto Protocol, signed in 1997, represents the first attempt to create a market to trade the major negative externalities produced by industrial society—greenhouse gases. Emissions trading has a long lineage going back to the 1960s. Looking at the problem of how to regulate overcrowded commercial radio waves, the American economist Ronald Coase (1960) suggested that frequency interference between radio stations could be reduced by defining clear property rights over specific radio frequencies. The logic was that broadcasters would want to pay for something that was previously free if it would guarantee that there would be no interference to their signal. Coase argued that this would create a system of "prevailing efficiency," whereby the party who could use the bandwidth most effectively (i.e. profitably) would ultimately end up paying the most for it.

Applying Coase's Theorem to waste water management in 1968, Canadian economist John Dales (1968) came up with a "cap-and-trade" system, whereby transferable pollution rights were allocated to market

Table 6.1 ***Taxes versus cap-and-trade***

	Tax	*Cap-and-trade*
Administration	Simple	Complex
Outcome	Uncertain	Certain
Price	Certain	Uncertain
Linkages	Hard to align	Easier to link
Flexibility	Very little	Built-in

participants up to a total quota of overall pollution that was deemed to be acceptable. Organizations that could easily reduce their pollution, he argued, would be incentivized to do so because they could sell their excess pollution rights to firms who were either less efficient or operating in a way that made reducing pollution very costly. Rather than forcing all organizations to reduce pollution by a set amount or in a set way, cap-and-trade systems allow individual organizations to respond in the way that is most effective for them, allowing overall reductions to be achieved for a lower overall cost.

Table 6.1 considers some of the advantages of cap-and-trade over traditional regulatory approaches, like simply levying a blanket tax. While taxes can be implemented simply by passing a law, their impact is uncertain. For example, evidence shows that while raising the price of gasoline reduces driving in the short term, levels tend to return to normal over time. In his most recent book, *Smart Solutions to Climate Change*, Bjorn Lomborg (2007) focuses on the most cost-effective ways to spend money to address climate change. His solutions include governments investing in research and development for new technologies, climate engineering and planting more trees. In cost-benefit terms he does not support an emissions tax, claiming that it would incur significant economic costs without achieving its stated goals of reducing emissions.

By contrast, cap-and-trade starts with the desired outcome, corresponding to an overall level of tolerable emissions, which is then allocated to users. This resonates with the broader preference of governance approaches to set targets but not prescribe how actors must achieve them. As Table 6.1 shows, cap-and-trade systems are also flexible because the amount of overall emissions permits in circulation can be raised or lowered. So, for example, the California cap-and-trade system that is due to come into force in 2011 is only intended to make up some 4 percent of the overall planned state reductions as part of the Clean Air Bill (signed by then governor Arnold Schwarzenegger in 2006), which aims to return emissions to 1990 levels by 2020. The plan is that it can be tightened up to achieve greater reductions if other measures fail, or the economy picks up.

Cap-and-trade systems are politically acceptable, because the price of carbon can be escalated gradually by slowly reducing the number of emissions permits in circulation. As the price increases, it gradually becomes rational for organizations to reduce their emissions as they develop alternative technologies. The political acceptance of cap-and-trade approaches to pollution control was marked by the passing of the Clean Air Act in 1990 in the USA (Environmental Protection Agency 1990). Created to control the industrial sulfur dioxide emissions responsible for acid rain, the Act established the first large-scale (national) market to trade atmospheric emissions, and had considerable success in driving a sharp decrease in emissions at relatively low cost.

The EU Emissions Trading Scheme devised at Kyoto is based on similar cap-and-trade principles, whereby a regulatory authority sets an overall cap on emissions and then allocates tradable permits to actors, which allow them to discharge a set quantity of emissions (Buckley *et al.* 2005). Begun in January 2005, the EU Emissions Trading Scheme is easily the most ambitious attempt to put the principles established at Kyoto into practice. A total of $92 billion of the $126 billion that the global carbon market was worth in 2008 was generated by the EU Emissions Trading Scheme. The principle of precedent use was employed to allocate free allowances to specific actors whose business is completely dependent upon producing emissions (for example, coal-fired power plants). If an individual organization exceeds its emissions allowance then it must buy additional quota, and vice-versa.

To date, the biggest problem has been the surplus of credits in the system, which has meant that carbon credits have remained too cheap. While it can be argued that prices need to start low in order to allow the development of alternatives to catch up, pricing should significantly alter the activities of the market participants otherwise they are failing to steer behavior. Of course, one of the key challenges to the EU Emissions Trading System was that compulsory monitoring of emissions was only implemented at the same time as the market itself, which meant that allocations were based on a lack of solid information concerning the actual emissions of different actors.

Contrasting this with the US scheme to reduce sulfur dioxide emissions in the 1990s, acid rain is caused by a single pollutant (sulfur dioxide), originating from a limited number of point sources in the energy sector (for example, coal-fired power stations). By contrast, greenhouse emissions comprise a number of gases, which are emitted by all sectors of the economy, making them far harder to regulate in a single market. There are no guarantees that a cap-and-trade scheme can simply be applied off the shelf to govern greenhouse gas emissions (Ellerman *et al.* 2000).

If Stern and others are to be believed, and emissions trading schemes are going to save the world, then it is important that different schemes are gradually integrated. Individual markets need to have border measures to prevent goods entering that do not comply with similar regulations, a problem known as carbon leakage. Japan's Voluntary Emissions Trading Scheme, launched in 2005, supports voluntary commitments by organizations to reduce emissions with subsidies and emissions trading. Participants of the Japanese scheme are a part of the Experimental Integrated Emissions Trading System (2008) and, when regional networks like the Western Climate Initiative in the USA and Canada come online there will be potential to link these schemes together.

The Kyoto Protocol also established two "baseline-and-credit" systems: Joint Implementation and the Clean Development Mechanism (CDM). Joint Implementation allows Annex I countries to offset their national emissions by investing in emissions reductions projects elsewhere, while the CDM allows Annex I countries to purchase credits that have been created by private organizations in the developing world. The CDM is intended to transfer clean technology and renewable energy systems to the developing world, by providing a revenue stream for investment in sustainability projects (Anderson and Richards 2001). Rather than capping emissions, baseline-and-credit systems allow organizations to emit pollutants up to a certain baseline. Baseline-and-credit systems differ from cap-and-trade in two important ways. First, rather than being allocated credits, organizations create them when their emissions fall below their respective baseline target. Second, rather than calculating the total emissions of an organization, baseline-and-credit systems calculate net emissions on a project-by-project basis (Buckley *et al.* 2005).

In order to take part in the CDM, individual countries must establish national accrediting authorities to certify that projects meet requirements. The most important of these are additionality, baseline, and sustainable development. The UNFCCC (2001: 3) defines the baseline as "the scenario that reasonably represents the anthropogenic emissions by sources of greenhouse gases that would occur in the absence of the proposed project activity." If the emissions of the project are below the baseline, then it can enter the CDM.

A number of criticisms have been leveled at the market mechanisms created by the Kyoto Protocol. The CDM involves a massive cast of public and private actors at local, national and global levels. Private investors are required to finance projects, developers are required to bring projects to market, NGOs are required to form the networks that link these actors together and spread information and know-how, and the UN has to run the accrediting bodies and regulate the CDM registry administrators and accountants (Boyd *et al.* 2007). Establishing a project is complex and time consuming, involving project design, validation, registration, monitoring, verification, certification, and the issuance of credits (Cozijnsen *et al.* 2007). Much of this effort is expended trying to create fungible units of carbon, so that each unit represents the same amount of carbon sequestration potential. The huge apparatus devoted to certifying projects is to ensure that emissions certificates produced by a hydro-electric dam project in Brazil are identical to those produced by a reforestation project in South Africa. The need for fungibility applies equally to any future global carbon market—a unit of carbon emitted in Guang Dong and traded on the Hang Seng must be substitutable for one produced in New Jersey and traded on the Dow Jones.

Originally, the CDM was intended to provide a fund for mitigation and adaptation in the developing world, but the way in which developed countries negotiated it meant that it ended up looking more like a fully fledged emission permits market (Bumpus and Liverman 2008). It is governed by international agencies based primarily in the developed world, while the networks of private consultants that emerged to verify and validate projects, who make large amounts of money out of the entire process, are drawn from the educated elites of the developed world. The majority of CDM projects have also come from areas of the world that are better equipped to negotiate the tortuous process of establishing and certifying projects (the World Bank estimates that 83 percent of CDM projects come from Asia). Marketing forest carbon in places like Mexico is hampered by a lack of institutional capacity in government and civil society, uncertainties in the international policy process, and the complexities of working with existing common property institutions (Corbera and Brown 2007). This highlights a core tension in market approaches, that in order to produce fungible units they must be disembedded from their social and ecological context. Research identifying CDM projects that have reforested areas by displacing subsistence farmers suggests that the system is geared more towards producing marketable products than sustainable development (Parreno 2007).

The market mechanisms created by the Kyoto Protocol have also been criticized for failing to change the behavior of emitters. The underlying logic of allowing developed countries to offset their emissions essentially allows them to continue polluting the atmosphere. Lohmann (2006) suggests that market-based trading mechanisms allow Annex I countries to continue with business as usual, preventing the kind of major changes to society that are required to move away from fossil fuel dependency. Worse, by simply paying developing countries to conserve their resources, mechanisms like the CDM ensure that the developing world remains underdeveloped (Bachram 2004). As a result, the CDM has been labeled a form of carbon colonialism, whereby the developed world simply exploits the carbon abatement potential of the developing world to maintain its standard of living (Harvey, F. 2007).

Finally, the complexities of creating fungible credits for exchange cast doubt over the ability of markets to deliver the necessary scale of change that is required to combat climate change. While the CDM had a market value of $24 billion in 2008 (Stokes *et al.* 2008), the World Bank estimates that developing countries will require $165 billion of investment in renewable energy in 2010, rising by 3 percent a year until 2030 (Boyd *et al.* 2007). The cost and complexity of establishing markets in tradable environmental goods casts doubts over the ability of these approaches to deliver the required quantity of investment quickly enough to help developing countries mitigate and adapt to climate change. That said, similar schemes are expected to form a key component of whatever agreement succeeds Kyoto, particularly through their extension to cover avoided deforestation as well as the creation of new carbon sinks, a possibility discussed in Case study 6.1.

Valuing the environment

Valuing the environment in financial terms highlights the potential economic costs of over-exploiting environmental resources, but can also

Case study 6.1

Post-Kyoto: the REDD schemes

Seventy percent of global forest carbon is located in countries which currently have high deforestation rates, which are defined as the loss of more than 0.22 percent of forest cover per year. According to the IPCC (2007), 1.6 billion tons of carbon were emitted annually in the 1990s due to tropical deforestation, constituting 20 percent of global emissions. The first scheme to reduce emissions from deforestation and degradation (REDD) was proposed to the UNFCCC in 2005 at the 11th Conference of the Parties in Montreal, and was soon joined by a further 19 governmental proposals and 14 non-governmental proposals in preparation for Copenhagen.

The REDD framework proposes to financially compensate developing countries for avoided deforestation and degradation, and is seen as a key component of the post-Kyoto framework to reduce global emissions and fund sustainable development. Popular schemes focus on reducing emissions from deforestation and degradation (REDD), while the most recent also look to enhance carbon stocks (REDD+). Most focus on above ground biomass (trees and vegetation), although below ground biomass (roots and leaf litter), soil carbon, or all of the above are scientifically justifiable, if harder to quantify in practice. There is a good rationale for starting with the simplest system, in order to enable developing countries to build capacity in carbon accounting practices, and then incorporate more complex elements like the enhancement of carbon stocks.

Most REDD proposals suggest that voluntary funds are used to pilot schemes and build capacity in the earlier stages, but few deny that only markets can provide the financial resources required to scale activities up to the global scale (Parker *et al.* 2008). Indeed, non-Annex I parties are leading the call for markets as they are aware of the shortcomings of current voluntary funding from the developed world, such as Official Development Assistance, which is insufficient and often tied to conditions. REDD would generate carbon credits that could be purchased by Annex I parties in exactly the same way as the credits currently produced by the CDM. If the problems of establishing fungibility of REDD credits are too great, then a market-linked mechanism may be established which trades REDD credits alongside other existing emissions credits, rather than in the same market.

As for the CDM, the job of enclosing carbon pools for REDD is not straightforward, requiring scientific bodies to define them, political organizations able to trade them, and someone to monitor this whole process. In terms of distributing money, most proposals simply assume that the benefits should go to the countries who are chopping down fewer trees, but this runs the risk of punishing countries with currently low rates of deforestation but high forest cover. In the worst case, REDD will provide an incentive for them to begin deforesting in order to be paid for subsequently stopping. In order to avoid such perverse outcomes a central distributive fund would be required, even if REDD operated as a direct market.

identify the actions where investment will produce the greatest good. In capitalist societies, money provides a common basis for comparison, and doing the "right" thing becomes instantly justifiable if it can be shown to be financially sensible. Responding in part to a political climate that has become increasingly led by economic considerations, a series of high-profile efforts have been made to demonstrate the economic worth of environmental goods that have in the past simply been used for free. The logic is that if these things can be financially valued, then they can be protected by being bought and sold in a market place. The final section of this chapter looks at three examples of how the environment has been financially valued: the Stern Review on the Economics of Climate Change, the McKinsey cost curve for climate change abatement, and the ecosystem service approach.

The Stern Review on the Economics of Climate Change

In 2005, the British chancellor of the exchequer, Gordon Brown, asked Nicholas Stern to review the economics of climate change to inform government policy. As former chief economist at the World Bank, Stern's appointment reflected the desire to engage an established and serious economist whose conclusions would carry weight beyond the environmental sphere. The review modeled a range of economic growth scenarios under the IPCC predictions for climate change, analyzing the costs and benefits of different degrees of political action to tackle the problem.

While considerable uncertainties surround climate change predictions, the report estimated that the overall costs and risks of climate change will be equivalent to losing between 5 and 20 percent of global GDP every year from now if no action is taken to reduce emissions (Stern *et al.* 2006). By contrast, the costs of reducing greenhouse gas emissions to avoid the worst impacts of climate change amount to around 1 per cent of global GDP each year. According to Stern's calculations, taking strong measures in the next 10–20 years to mitigate climate change will produce net global benefits of $2.5 trillion.

Stern suggested that three mechanisms can deliver the necessary reductions, all of which exist currently:

Emissions trading: expanding and linking the growing number of emissions trading schemes around the world and channeling revenues to support the transition to low-carbon development paths in the developing world.

Technology: increasing cooperation in developing new technology, specifically in the development and deployment of low-carbon technologies.

Reducing deforestation: using large-scale international pilot programs to explore the best ways to reduce deforestation, which contributes more to global emissions each year than the transport sector.

A key criticism made of Stern's calculations was that he did not discount the cost of future impacts sufficiently, and thus over-emphasized the potential economic costs of climate change. Stern has replied that discounting makes little sense in philosophical terms, as it may well be a cost to oneself which will be incurred in the future. Discounting also contravenes the demands of sustainability to consider the welfare of future generations. At its most basic, discounting the costs of future impacts effectively works against taking the long-term view. That said, even applying more conventional (i.e. higher) discounting rates to Stern's calculations produces a similar conclusion—mitigating now to prevent severe climate change is more cost-effective than adapting to it later. Cost-benefit analysis is discussed further in Analytics of governance 6.1.

While Stern's review has been criticized for some of its methods, it is generally accepted that his conclusions are sound (Arrow 2007). In providing strong evidence for the costs of inaction, the review counters the argument beloved of climate skeptics that mitigation now will be more costly than simply adapting to change in the future. But perhaps the most important impact of the Stern Review has been to make people think of climate change in economic rather than purely scientific terms. In a famous passage the review refers to climate change as a "market failure"—in other words, a failure to correctly value the resources that we use. While this contrasts with Mike Hulme's (2009: 310) assessment noted in Chapter 1 that climate change is a "crisis of governance . . . [not] a crisis of the environment or a failure of the market," it is representative of the conviction among economists that markets are not inherently bad for the environment, but can help if designed correctly. Stern has helped establish climate mitigation as a major consideration for governments around the world.

The McKinsey cost curve for climate change abatement

Cost-benefit analysis has also been used to identify activities with the greatest potential for carbon abatement (emissions reductions). Private

Analytics of governance 6.1

Cost-benefit analysis

Cost-benefit analysis "involves the monetization of all of the costs and benefits of a proposed policy, plan or project (including alternatives) and the assessment of the resultant net benefits over a given time horizon" (Petts 1999: 37). It is used as a decision-making tool to determine whether a project should go ahead by identifying all of the impacts and effects, assigning monetary value to them, aggregating them and calculating whether the benefits outweigh the costs. Cost-benefit analysis primarily concerns evaluation rather than prediction, producing common monetary measures to compare policy or project alternatives in a robust and transparent way.

While the benefits of a policy or project are often felt immediately, many of the costs will be incurred in the future. So, for example, while a nuclear power plant will produce energy two or three years after the start of construction, the cost of having to deal with the resulting nuclear waste will grow over time, culminating with the decommissioning of the power plant itself in about 50 years time. One of the most controversial aspects of cost-benefit analysis is that it applies a future discounting rate, which means that a cost or benefit now has more weight than a cost or benefit further down the line. Discounting rests on the assumption that people in the future will be better equipped to deal with potential costs, by, for example, being wealthier, or having more advanced technology. In terms of nuclear energy, we had better hope this is the case. The world's 441 functioning nuclear power plants produce a combined total of 13,000 tons of highly radioactive waste per year, but there are currently no permanent stores in which to entomb it (Weisman 2007).

Humans tend to prefer short-term gains at the expense of long-term costs, a trait that behavioral economists call "hyperbolic discounting," whereby future costs are literally discounted in the calculations people make about how to act in the present. The tendency is said to be hyperbolic because it becomes more pronounced the farther away the problems are perceived to be. Hyperbolic discounting presents serious problems for decision-makers, because many of the most severe effects of climate change will not be felt for 50 years or more, making it extremely hard to generate support for mitigation that may require sacrifices now.

Environmental philosophers have also attacked cost-benefit analysis for failing to take account of previous actions and decisions. Project or policy evaluation starts from "year zero," ignoring previous decisions or wider cultural preferences, operating as if decisions are taken in a historical and political vacuum (O'Neill 2007: 87). The tendency of market-based approaches to disembed decision-making from its social context runs counter to the principles of sustainability, which emphasize the involvement of communities in locally appropriate action.

By contrast, climate scientists have criticized cost-benefit analysis because it is unable to deal with the future accurately. The calculation of costs and benefits involves simply extrapolating current trends, assuming that social and environmental changes will follow a broadly similar pattern in the future. This is of course far from guaranteed in the context of climate change, which is characterized by non-linear changes and tipping points. Although it forms the main tool which policy-makers use to assess their responses to climate change, cost-benefit analysis breaks down entirely under non-linear conditions. By starting from a desirable future point, the 2°C guardrail advocated by the international scientific community explicitly tries to create a window of predictability within which non-linear changes to the climate are less likely, and traditional models of decision support, like cost-benefit analysis, can function (Kates *et al.* 2001).

consultancy McKinsey and Company (2009) produced one such analysis for the Swedish energy company Vattenfall AB, which ranked different abatement activities in terms of their overall costs and benefits. Figure 6.1 shows the resulting cost curve for carbon abatement, with the potential size of each abatement measure on the horizontal axis in gigatons of emissions ($GtCO_2e$) per year, and the net cost of that measure, in euros per ton of avoided greenhouse gas emissions, on the vertical axis. The curve is based upon the maximum possible savings for each abatement measure in the 20 years up to 2030, if currently available technical solutions are pursued as aggressively as possible.

Figure 6.1 also shows how the global emissions reductions associated with abatement activities translate into lower atmospheric greenhouse gas concentrations, marked at the 550, 450 and 400ppm levels along the x-axis in the middle of the diagram. Best estimates suggest that reducing global emissions by 26 Gt of CO_2e per annum would stabilize the atmosphere at 450 ppm of CO_2e. For context, the IPCC estimates that a greenhouse gas concentration of 450 ppm of CO_2e gives a 50 percent probability that the eventual global temperature rise will exceed 2°C. As discussed in Chapter 1, 2°C is seen as a critical guardrail, because above this level climate impacts become very severe. The “26” in the shaded circle on the x-axis of Figure 6.1 indicates that to achieve a reduction of this magnitude in the next 20 years would require all abatement activities up to the cost of €40 per ton of CO_2e to be vigorously pursued. To achieve the entire 38 Gt of abatements that are possible would require $490 billion of investment per year by 2020 and $860 billion by 2030.

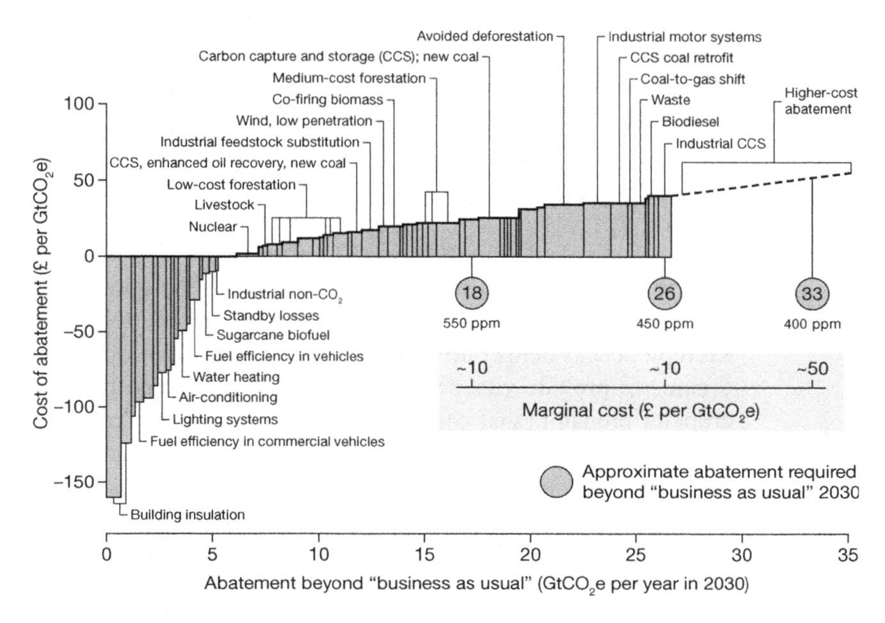

Figure 6.1 ***Global cost curve for greenhouse gas abatement measures beyond "business as usual"***

Source: adapted from Enkvist *et al.* 2007.

Drilling down into the global picture reveals that the biggest sectors in terms of abatement potential are power generation (26 percent) and forestry (21 percent). The bulk of investment (approximately 75 percent) is required in the power, transport and buildings sectors, and, accordingly, 55 percent of the overall investment will be required in China, North America, and Western Europe. Conversely, 70 percent of the actual abatement opportunities are in the developing world, and they cost considerably less to achieve. The lowest hanging fruit is the forestry sector, which requires less than 5 percent of the overall investment, but makes up over 20 percent of the abatement potential. The curve powerfully demonstrates that the easiest opportunities to avert damaging global climate change lie in the developing world.

The McKinsey cost curve indicates that pursuing the most economically efficient abatement opportunities up to a cost of €40 per ton of CO_2e would cost between €200 and €350 billion per year, less than 1 percent of the forecasted global GDP for 2030. Echoing the recommendations of the Stern Report, McKinsey notes that this makes mitigation activities considerably cheaper than the IPCC's (2007) best estimates of adaptation costs, which stand at around 5 percent of global GDP. Like Stern, it considers these activities to be "within the long-term capacity of global financial markets."

The ecosystem service approach

Rather than calculating the costs of various lines of action or inaction to society, the ecosystem service approach values the goods and services that natural systems and biological diversity provide for humans. For example, ecosystems purify water and clean the atmosphere, while organisms provide vital services, like the bees that pollinate commercial crops or predators that control natural pests. Robert Costanza's (1997) famous paper in the journal *Nature* estimated the value of global ecosystem services to be $33 trillion per year. Obviously total values are not very helpful here—the Earth's atmosphere is literally invaluable to us as we would die without it. Rather, valuations of ecosystem services help us to understand the impacts of marginal change, calculating, for example, the financial costs of negative health impacts associated with a 5 percent increase in air pollution, rather than the total cost of air pollution or the value of clean air.

There are four main categories of ecosystem services (de Groot *et al.* 2002):

Provisioning services: ecosystem services that provide direct goods, such as food.

Regulating services: ecosystem services that condition the environment and maintain its health, such as water and air quality.

Cultural services: ecosystem services that provide non-material benefits, such as recreation.

Supporting services: ecosystem services that underpin the production of the other three ecosystem services, like soil formation.

In order to aid decision-makers, ecological economists have attempted to place financial values on the services that the environment provides us with. Because the ecosystem services approach produces financial values, it allows direct comparisons between the costs and benefits of different types of decisions. Of course, environmental impacts, services and goods do not simply come with a price tag already on them and there are a number of economic valuation methods that can be used to

Key debate 6.1

The ethics of financial valuation

There are a number of issues with hedonic valuation. Sometimes people's enjoyment of a resource can't be expressed in financial terms, while the amount that people are willing to pay will reflect not only the worth that they personally attach to the service, but how wealthy they are. This means that a forest used by a few rich people may be valued far more highly in financial terms than one accessed by many poorer people, despite the fact that one resource has far greater use value. Further, valuing the environment purely in terms of its utility to humans means that services that are not currently of use but might be in the future, or that people are not aware of, are not valued. The flipside is that environmental resources that are used frequently have more value than those that are not. So for example, the London Tree Officers Association recently valued a single plane tree in Mayfair at three quarters of a million pounds, based on its enhancement of already exhorbitant property prices and its sheer visibility.

There is some part of most people that feels an aversion to valuing the environment in financial terms. How can we place a dollar sign on the spiritual uplift that accompanies a beautiful sunset, or balance the implications of driving a species extinct against the value of an untapped oil reserve? One of the most vocal critics of financial valuation, Mark Sagoff (2004), argues that care for the environment is an ethic in itself, which is actively undermined as soon as environmentally friendly actions are reduced to financial transactions. Markets reward people to behave badly in order that they can then be paid more to behave well. As he says, "the thing becomes completely corrupt as every single person who might be able to control carbon by farting less demands a credit" (quoted in Jenkins 2008).

A leaked memo in 1991 from Harvard economist Lawrence Summers, then chief economist at the World Bank, gave an infamous insight into the problem of applying economic logic to environmental problems (Harvey, D. 1996). The memo began, "Just between you and me, shouldn't the World Bank be encouraging the migration of dirty industries to the Less Developed Countries?" It went on to argue that because the costs of health-impairing pollution depend on lost earnings from increased morbidity and mortality, rich countries should dump toxic waste in the lowest wage country. The costs of pollution are likely to be non-linear as the initial increments of pollution have a low cost, meaning that the air quality of non-industrialized countries is "inefficiently" low. He went on to argue that it's "lamentable" that so many air polluting industries are "non-tradable," as "externalizing" health costs from the world's rich to the world's poor would raise their income.

The Washington office of Greenpeace copied the memo to environmental groups around the world, prompting general disbelief and outrage. Brazil's secretary of environment called it "perfectly logical but totally insane," while the *Economist* magazine hailed the "impeccable economics" while warning that we need to "save planet Earth from economists." While this form of toxic imperialism showed up the glaring social and environmental flaws of neoclassical economics, Summers went on to become President Clinton's under-secretary of state for trade. A similar logic underpins all economic approaches to the environment. For adaptation assessments, the IPCC values a statistical life in the developed world at $5 million, compared to one in the developing world at $0.5 million.

Costanza, who first estimated the worth of the world's ecosystem services, makes an interesting comparison with the way in which the worth of human lives is traded off against the cost of installing extra safety measures on highways to reduce fatalities (Jenkins 2008). As he states, this kind of calculation values a statistical life, rather than a particular person. Further, in some instances there may be ethical reasons to destroy the environment, for example if it is the only way to feed people. The fear remains, however, that upon entering the decision-making system, financial values are taken to be completely representative of the worth of an environmental resource and other considerations are simply lost.

value the environment, each with merits and weaknesses. Most involve some form of "hedonic pricing," whereby proxy prices are generated by asking people to express individual preferences in monetary terms. For example, willingness-to-pay might ask users of a forest to estimate how much they would be willing to pay in order to use the car park from which they access it. This gives a proxy price for the worth of the forest in terms of the services it provides. Some of the tensions surrounding the financial valuation of ecosystem services are discussed in Key debate 6.1.

Cognizant of the considerable methodological flaws in his work, Costanza stated that his primary aim was to highlight the potential value of the Earth's ecosystems in order to raise awareness. The ecosystem service approach has certainly gained serious ground in the last 10 years as the basis for environmental decision-making. The Millennium Ecosystem Assessment, discussed in Chapter 3, represents an attempt to measure the state of the world's ecosystems and provide a scientific basis for the ecosystem services approach. Ecosystem services are recognized by the Convention on Biological Diversity, which provides 12 principles and five points of operational guidance. It has also been used in the USA and in the EU to drive more environmentally focused agricultural subsidies.

In valuing environmental services, the ecosystem services approach performs a similar role to the Stern Review, highlighting the importance of environmental goods that are otherwise simply ignored in decision-making. The difference is that whereas Stern's calculations were intended to stimulate and steer collective political action to address climate change at the national and global level, the ecosystem service approach is intended to support specific decisions right down to the local scale of development control.

Conclusions

To those charged with addressing environmental problems, market principles hold considerable appeal. Valuation exercises like those described above show that the environment has considerable financial worth, which lends it greater weight in decision-making. In terms of governance, the question becomes how to insert these values into the market system that underpins capitalist economies. On the other hand, there are a number of critiques of market approaches that question their ability to deliver the rapid transformations required to address climate change. Table 6.2 summarizes the strengths and weaknesses associated with market governance.

The key appeal of market approaches is that they promise efficiencies in the way that environmental problems are addressed: efficiency of decision-making through the laws of supply and demand, which pool

Table 6.2 ***Strengths and weaknesses associated with market governance***

Strengths	*Weaknesses*
Efficient pooling of knowledge under conditions of uncertainty	Impossible to capture all aspects of the environment in monetary terms
State involvement is lower and thus schemes are cheaper	Practical problems with enclosing environmental resources that make it hard to create fungible commodities
Can bring potentially huge resources to bear upon a problem	Can be hard to distribute resources fairly and selectively
Prices can be manipulated in order to steer economic activity towards environmentally desirable outcomes	Markets can be captured by business interests simply maintaining existing inequalities
Recognizes complexity of the real world	Difficulty of market design, leakage, etc.

collective knowledge; efficiency for governments, whose role is simply to regulate the market; efficiency of steering, by manipulating prices and incentives; and efficiency of action, by bringing massive amounts of resources to bear upon a problem in a fairly short timeframe. To a large extent, the weaknesses simply question each of these supposed efficiencies. Philosophers question the ability of environmental services to be captured by financial valuations, and there is no doubt that the process of enclosing many environmental goods to make them tradable in a market is hugely complex, expensive, and time-consuming. While markets generate wealth, they are notorious for generating inequalities and reinforcing the status quo, and do not always direct money to the places and people who need it most. Part of this problem involves the challenges of creating and regulating new markets, which must be incentivized generously enough to be acceptable to those taking part, but also be stringent enough to actually change their behavior. The challenge at the global level is to avoid leakage by creating markets that are international in order to avoid companies simply relocating outside of a market area.

Many of the factors in Table 6.2 illustrate that markets do not operate in a vacuum, but within parameters set by the state. In the environmental field, markets like those created by the Kyoto Protocol require an army of NGO and corporate institutions to implement them. Governments play a key role, with the ability to pass laws that literally create new markets overnight (like those associated with the Kyoto Protocol). From recycling to carbon trading, the state has the ultimate power to create and destroy entire industries, and, as the proposed REDD schemes suggest, the question is rarely "market or no market," but rather what role markets should play as part of a mix of governance approaches (a point that is returned to in the final chapter).

While neoliberals tend to attribute to markets an almost mystical ability to simply spring up in the absence of constraining regulations, as if they were hardwired into instinctive human behavior, most experiments in creating markets suggests that they are actually rather fragile things that can survive only in a highly protective womb of learnt cultural behaviors and legal frameworks. As Andrew Gamble (1992) argues in *The Free Economy and the Strong State*, free markets require strong government institutions to prevent monopolies, encourage competition, and deal with unionized labor if they are to function correctly. Scholars like Becky Mansfield (2006) come to similar conclusions concerning her study of North Pacific fisheries—markets require regulations in order to function. This brings us neatly on to the topic of the next chapter—the role of the state in steering economic development towards a sustainable transition.

Questions

- What are the key institutional requirements for a global carbon market?
- Markets are criticized for reinforcing economic inequalities between rich and poor. Does this matter in relation to their ability to address environmental problems?

Key readings

- Bumpus, A. and Liverman, D. (2008) "Accumulation by decarbonisation and the governance of carbon offsets," *Economic Geography*, 84: 127–55.
- Costanza, R. d'Arge, de Groot, R., Farber, S., Grasso, M., Hannon, B., Limburg, K., Naeem, S., O'Neil, R. V., Paruelo, J., Raskin, R. G., Sutton, P. and van den Belt, M. (1997) "The value of the world's ecosystem services and natural capital," *Nature*, 387: 253–60.
- Stripple, J. and Lövbrand, E. (2010) "Carbon market governance beyond the public–private divide," in F. Biermann, P. Pattberg and F. Zelli (eds) *Global Climate Governance Beyond 2012*, Cambridge: Cambridge University Press, 165–82.

Links

- www.storyofstuff.com/capandtrade/. Annie Leonard's engaging cartoon introduction to the leading solution to climate change.
- www.pointcarbon.com. News about global carbon markets.
- www.mckinsey.com/clientservice/sustainability/pathways_low_carbon_economy.asp. Lots of cost curves and corporate low-carbon strategy documents from a leading global business consultancy.

7 Transition management

Intended learning outcomes

At the end of this chapter you will be able to:

- **Articulate what a technological transition is and how it applies to sustainability.**
- **Understand the relationship between society and technology.**
- **Identify the key characteristics of transition management as a distinct approach to environmental governance.**
- **Appreciate the strengths and weaknesses of transition management.**

Introduction

> The stone-age didn't end because we ran out of stones.
>
> (Sheik Ahmed Zaki Yamani, 1973)

Sheik Yamani's remark, made at the height of the first oil crisis, implies that the world does not have to wait for oil to run out before it embraces the alternatives. Low carbon technologies promise to square the circle of environment and development by decoupling economic growth from carbon emissions. As the practice of directing technological developments in society, transition management has obvious relevance to the challenges of transitioning to a low carbon economy, and focuses strongly on the steering dimension of governance.

This chapter explores the kinds of systemic transformations that are required to achieve a transition to a low carbon economy. It begins by considering the work of scholars, who have explored how isolated technological innovations spread through society to create a so-called technological transition. A series of case studies, including smart grids, cycling, and electric cars, is used to demonstrate the importance of social, political, and economic factors in explaining why technologies succeed or fail. The final part of the chapter assesses how transition management has fared in practice, drawing on the example of energy policy in the Netherlands, before concluding with a discussion of its strengths and weaknesses.

Technological transitions

From the four-stroke combustion engine to the internet, the development of what we know as modern civilization has been punctuated by a series of major technological innovations. In each case, new technology has been developed, trialed and rolled out to wider society, often at great cost. Sometimes a new technology directly replaces a predecessor, as the railways made the canals of Britain redundant in the nineteenth century even before the national network of canals had been completed. Achieving complete broadband internet coverage in many developed countries is currently requiring significant upgrades to the communications infrastructure in the shape of laying fiberoptic cables. In the developing world, technologies like mobile phones are spreading without there ever having been a landline network. In relation to sustainability, low carbon technologies might allow the developing world to leapfrog the older, dirty technologies that were used in the developed world. Considering technological transitions means studying the way in which these transitions take place, focusing on how technological innovations occur and are subsequently incorporated into society.

Figure 7.1 shows a stylized transition that describes how a technological innovation spreads though society over time. The diffusion of a new technology passes through a series of phases, from inception, through a break-out period, to dominance. The key questions that concern transition are what conditions encourage innovation, why some innovations break out and others don't, and how break-out innovations go on to become ubiquitous.

Dutch scholars have developed the concept of a technological transition to understand the process of innovation and diffusion. For them, sets of rules embedded in institutions and infrastructures generate specific technological trajectories, which are often embedded in communities of engineers or scientists searching for solutions to a similar problem (Rip and Kemp 1998). These rules, and the communities that they bring into being, form what is known as a regime. The regime is the level at which

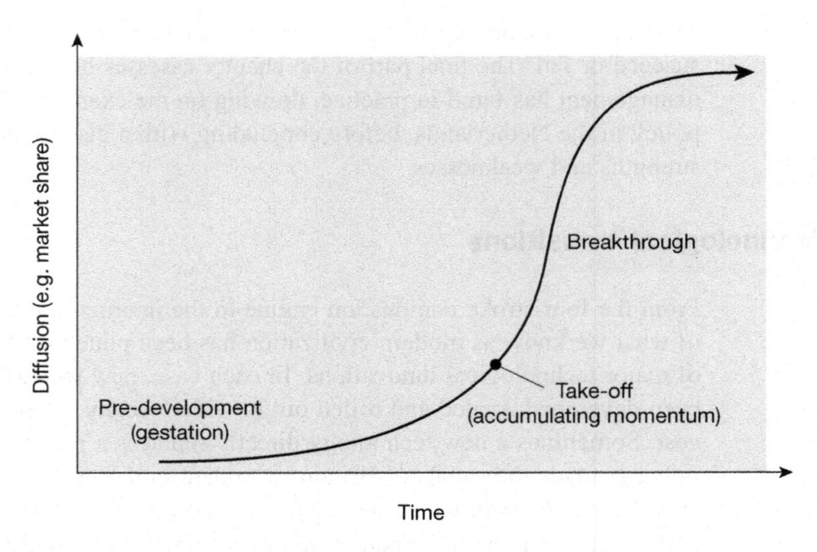

Figure 7.1 ***Stylized shape of transitions***
Source: adapted from Rotmans *et al.* 2001.

the basic functions of society are performed, for example maintaining power lines and substations in order to deliver energy. In turn, the regime is embedded in a wider socio-technical landscape, which is not unlike the order of meta-governance discussed in Chapter 2, constituting the wider political and cultural environment in which transition occurs. In relation to energy this would include things like the type, volume and distribution of energy resources, the wider policy-making agenda, and the cultural values and principles that relate to renewable energy (Steward 2008).

Drawing on insights from evolutionary economics, the transition approach sees innovations as competing with one another in the market place, with successful ones spreading into the wider regime and unsuccessful ones dying out. While incremental change is always taking place within regimes, radical change tends to originate in niches, which are protected environments in which innovations emerge and are tested (Geels 2002). Niches form incubation rooms where unique combinations of expertise and resources are available that provide the seeds for change (Kemp *et al.* 2001a). In any regime a number of niches may exist, generating a range of innovations and alternatives to the dominant way of doing things. Because radical innovations may be commercially unviable at first, "the creation of niches by social and political networks is critical to protect them from the constraints of the regime" (Hoogma *et al.* 2002: 25). A key role for the state is to build relations "between actors to support the innovation in very specific time and space contexts" (Beveridge and Guy 2005: 675), sheltering innovations from wider political and economic pressures, often through subsidies and tax breaks.

Figure 7.2 shows how the three levels of niches, regimes, and landscapes work together to create a transition. Put simply, niches are where innovation occurs, the regime is where selection occurs, and the landscape forms the broader context within which these processes take place. It is worth noting that this multi-level framework is heuristic, not ontological. In other words, it is not suggesting that the world is actually comprised of niches, regimes and so on, but rather that these are useful analytic categories that allow us to explain how technological change occurs.

As Figure 7.2 shows, this process is rarely revolutionary, but takes place through a series of adaptations over time. As niches accumulate, they begin linking together and break into the socio-technical (ST) regime,

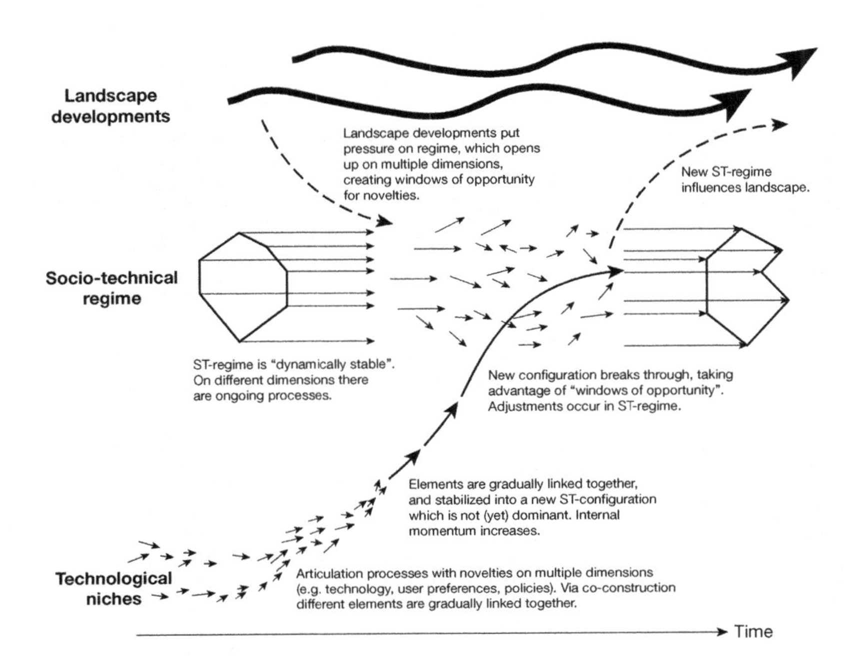

Figure 7.2 ***A dynamic multi-level perspective on system change***

Source: adapted from Geels 2004: 915.

destabilizing it until a new configuration of technologies becomes established. The regime is only ever in a dynamically stable configuration because it is in a constant reciprocal relationship with the landscape level, which creates windows of opportunity for change through cultural shifts, policy changes and so on. In showing how steam replaced sail ships, Geels (2002: 1262) traces the early stages of the first steamboat experiments, showing how they broke out of their niche when "ongoing processes at the levels of regime and landscape created a 'window of opportunity.'" Within this model of system change, old and new technologies usually co-exist for a period of time before one completely supersedes the other.

Geels *et al.* (2008: 7) identify six characteristics of technological transitions:

Transitions are co-evolutionary and multi-dimensional. The adoption and spread of new technologies is dependent upon both technical and social factors. For example, innovations in ICT are driving many erstwhile office-based workers to operate from home, which in turn has driven the expansion of fiberoptic infrastructure to deliver high-speed internet to residential areas. In such a way, technology and society are transformed at the same time.

Multiple actors are involved. By their very nature, system changes involve most social groups and stakeholders, including firms, policy-makers, consumers, suppliers, distribution and retail chains, civil society and NGOs. Because transitions are prompted by innovations, actors operating outside mainstream society often play an important role.

Transitions take place at multiple levels. System change typically involves interactions between processes at different levels, whereby change is transmitted from the niche to the landscape.

System changes are radical. While shifts from one system to another tend to be gradual rather than sudden, they eventually result in radical change.

Transitions are long-term. Transitions take several decades to complete.

The rate of change is non-linear. The changes during a transition are not constant, but vary over time, relating to the transition stages shown in Figure 7.2.

Transition and sustainability

The idea of low carbon economy is becoming commonplace in policy, and involves steering activity towards more sustainable industries. For example, UNEP's Green Economy Initiative is designed to assist governments in encouraging more environmentally friendly industries. The logic is sound—many of the activities that are required to make society more sustainable are highly labor intensive, and thus encouraging low carbon industry should stimulate job creation (the primary concern of almost all political leaders in democracies). For example, retrofitting existing housing to be more energy efficient, installing solar panels, or modifying cars to take biodiesel cannot be done by machines in factories, but requires an army of skilled workers. Many of the challenges of climate change involve large-scale technological transitions, from petrol to electric cars, or from coal-fired power stations to renewable energy, which will require massive changes to infrastructure and the way in which we live.

The question of how many jobs will be created by a green economy is critical in justifying decisions to redirect investment. This has become something of a political football, with advocates emphasizing that new, sustainable (in every sense) jobs will be created, and its detractors pointing to the jobs that will be lost by forcing polluting industries either to relocate or go out of business. Proposition 23, California's recent vote on whether to suspend its Clean Air Bill until state unemployment falls from over 10 percent to under 5.5 percent, has been fought almost exclusively over the question of whether the CleanTech sector can replace the jobs that might be lost by oil refineries leaving the state.

Studies of the German experience, where the federal government's Integrated Energy and Climate Program was implemented in 2007, suggest that 500,000 jobs may be created by 2020. Further, if the so-called Meseberg Program can solidify Germany's position as a global leader in renewable technology then a further 1 million jobs may come into being (German Advisory Council on Global Change 2009). The situation in the USA is more hypothetical, but studies suggest that investing comparable amounts in energy efficiency and renewables as the government currently spends subsidizing fossil fuel related industry would create approximately 20 percent more jobs (Houser *et al.* 2009, Pollin *et al.* 2008). There is no telling what will happen unless such policies are trialed, but this entails a considerable degree of political risk.

It was in this context that the 2008–9 global financial crisis was hailed by many as an opportunity to put the low carbon agenda into practice (Stern 2009). Not only did the crisis dent faith in the current system even to deliver economic growth, but it offered an opportunity to steer the economy onto a more sustainable path by supporting green industries with government money. Comparisons were quickly made with the New Deal offered by the US government in the wake of the Great Depression of the 1930s, which used massive public investment in infrastructure and social programs to generate employment. The double crisis of climate and finance offered the opportunity to create a global "Green New Deal," which would see governments matching their political rhetoric with economic investment in environmental sustainability (Leichenko *et al.* 2010). In the USA, President Obama's Green New Deal sought to use stimulus packages to redirect the economy towards more sustainable activities.

Figure 7.3 shows the percentage of gross domestic product that was invested in economic stimulus packages by key members of the G20 (the countries with the 20 largest national economies), and the proportion of this money that was targeted at green industry. UNEP set a target that at least 1 percent of GDP should be invested in green industry, but most countries failed to meet this. Globally, the average proportion of the overall stimulus package directed towards green investments was 15 percent, most of which went into rail and waste. China and South Korea stand out, investing 28 percent and 80 percent respectively, which compares to 12 percent for the USA. There is simply no doubt that green investment is being taken far more seriously in places like Germany and China than elsewhere.

There is an increasing range of technologies that support the goals of sustainability. Cradle-to-grave design takes into account the entire life-cycle of a product, including running and disposal costs, while biomimicry seeks to apply design principles from nature to human products (Webster and Johnson 2008). In proposing preventative rather than remedial environmental protection, these technologies address the causes of environmental pollution rather than treating their symptoms. Case study 7.1 discusses industrial ecology, which aims to maximize the way in which waste products can be reused.

Because existing technologies are already established they enjoy an inbuilt advantage over new ones, called lock-in. People own petrol cars

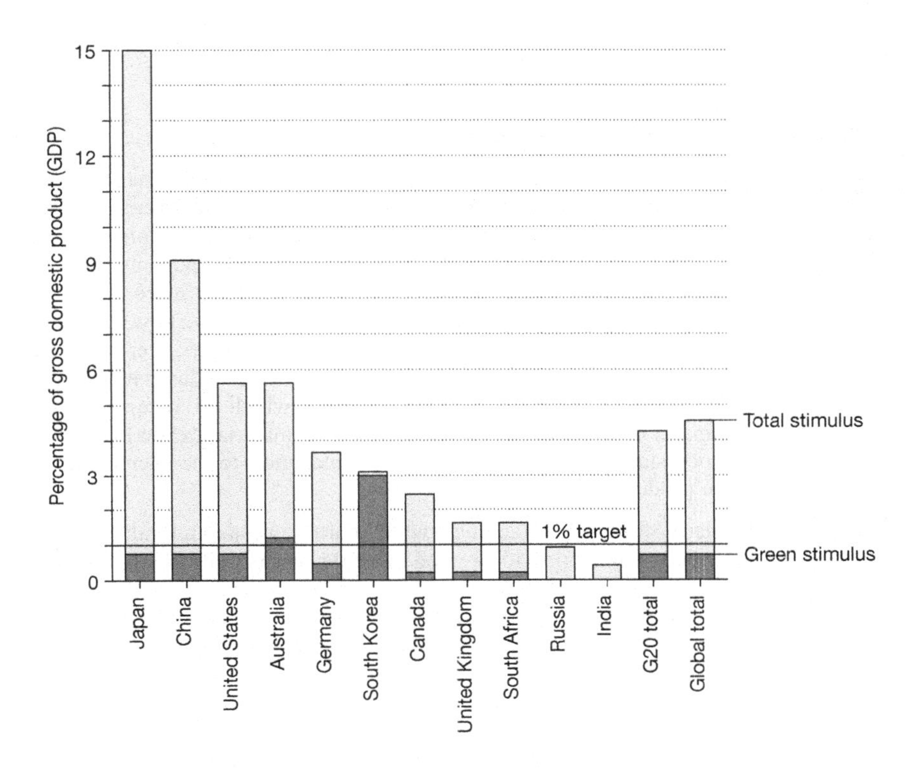

Figure 7.3 ***G20 green investments***
Source: adapted from Barbier 2010.

and understand their limits, mechanics can mend them, companies can insure them and there is a global infrastructure of filling stations. Not only is society locked in to existing technologies, but it is also threatened by the new. As Machiavelli (1992: 17, quoted in Lessig 2001: 6) stated almost 500 years ago, "innovation makes enemies of all those who prospered under the old regime, and only lukewarm support is forthcoming from those who would prosper under the new." Whether we are talking about electric or hydrogen powered vehicles, the successor to the car that we currently know will have to overcome social, technical, political, economic, and financial lock-ins to oil-powered transport. Case study 7.2 explores how a fundamental aspect of modern life like power supply presents substantial challenges to transition.

Case study 7.1

Industrial ecology

Industrial ecology is based on the idea that industry is a part of, rather than separate from, the biosphere. It aims to redesign manufacturing processes to behave more like ecosystems, whereby the waste of one process becomes the resource input for another. As two of its earliest proponents state, "materials in an ideal industrial ecosystem are not depleted any more than those in a biological one are; a chunk of steel could potentially show up one year in a tin can, the next year in an automobile and 10 years later in the skeleton of a building" (Frosch and Gallopoulos 1989: 2). Industrial ecology has developed tools like life-cycle assessment, which try to capture all the impacts of an industrial process, from raw material extraction, material processing, manufacture, use, and maintenance, to the eventual disposal of the product.

Energy-from-waste facilities, which burn domestic and industrial refuse to produce power, provide a good example of this philosophy in practice. As the amounts of waste produced by humans increases, simply burying it in landfill is becoming impractical and expensive. Energy-from-waste facilities create a closed-loop system by using waste as a resource, burning rubbish that cannot be separated for recycling or composting under highly controlled conditions to generate energy. The resulting ash passes through a handling system where metals are extracted, and is then sent for recycling within the construction industry and/or disposal. Hot gases produced in the combustion process are cooled in a boiler to produce steam, which then enters a turbo-generator to generate electricity for export to the local grid.

Opened in 1988, Vancouver's energy-from-waste facility has played an essential role in the region's solid waste management system, ensuring that garbage is managed in an environmentally safe manner. The facility is operated and maintained by Covanta Burnaby Renewable Energy Inc., and is located in the commercial/industrial area of south Burnaby. It is responsible for the disposal of over 20 percent of the region's waste. Each year the facility turns 280,000 tons of garbage into steam and electricity. The steam is sold to a paper recycling facility, while enough electricity is sold to BC Hydro to power 12,300 homes. The facility generates $10 million in annual revenues from electricity and steam sales alone.

Given these challenges, transition management appeals to those charged with making our society more sustainable. Transition management builds upon technological transition studies to distil lessons for directing long-term change in large socio-technical systems for sustainability. Financial and regulatory pressures are exerted on the existing system, while granting tax breaks and funding for research and the development of experimental projects (Geels *et al.* 2008). This may include deliberative processes, whereby stakeholders in a sector come together to dream up innovations, or funded intermediaries who act to translate research into policy. Meadowcroft (2009) identifies six aspects of

Case study 7.2

Smart grids

Traditionally electricity is generated by burning coal in massive, centralized power plants, the power then being distributed via a national grid comprised of substations and pylons. The system is responsive to fluctuating usage patterns, in that more energy can be produced at the times of the day when it is needed most (simply by cranking up the heat). By contrast, renewable energy provides what is called "lumpy," or uneven, supply, because wind power tends to be most available at night (when the wind blows), and solar power is dependent on the sun. Uneven supply creates mismatches between supply and demand, a problem exacerbated by the surprising fact that modern society lacks an effective way to store electricity. The idea behind storage heaters, which use electricity at night, would need to be extended to other technologies, and price tariffs would be required to heavily subsidize nocturnal use and punish daytime use. For these technologies to stand a chance of being successful, people would need to understand and buy in to new ways of living.

On the other hand, substantial infrastructure is required to move coal to the power stations (usually railroads), and to subsequently transport electricity around the country (substations, pylons and so forth). The fact that renewable power production would be highly dispersed, with lots of small points of production rather than a few massive installations, is an advantage in principle. Getting our energy from wind turbines, hydro-electric power dams and solar arrays would make society far more resilient to crises. Energy would be produced locally rather than imported from unstable areas of the world, operational problems would affect fewer people at once, and the grid would be supplied by a diversity of sources, meaning that if one failed another could step in.

But in the short term even this advantage is a barrier, because it requires "super-smart grids," capable of balancing fluctuating supply and demand from multiple sources. This represents a very different type of power grid to the one that currently exists. Feed-in tariffs can be used to promote small-scale renewable energy production, but the financial and human resources required to re-engineer the power grid represents a major bottleneck for energy transition.

transition management that make it a potentially useful mode of governance for achieving sustainability:

It makes the future clearer in current decisions. Generating multiple possible pathways and considering their viability and/or desirability over relatively long time periods allows sustainable futures to be incorporated into current decision-making.

It transforms established practices. The concept of the regime captures the importance of changing accepted ways of doing things and social behaviors in order to generate change.

It develops iterative processes that constantly self-assess and re-adjust. Actors come together to consider new solutions to specific production or consumption problems, and operate in an interactive and iterative way to address them.

It links technological and social change. As the issue of lock-in suggests, new technologies require social change in order to be accepted.

It emphasizes learning by doing. Transition management is experimental, in that it advocates trying things out in the real world and learning from them.

It encourages a diversity of approaches rather than a single, centralized plan. The concept of niche innovation generates multiple approaches, which are then selected for or against through the pressures of the wider environment. This diversity is more suited to the complexity and local requirements of sustainability than a single centralized plan.

Of course, one of the key differences between the kind of historical transitions that form the focus of the technological transitions literature and sustainability transitions is that the former are often the outcome of historically contingent processes, whereas the latter are purposefully steered (Hodson *et al.* 2008). The California Hydrogen Highway represents a directed attempt to shift from one technology (gasoline cars) to another (hydrogen cars). In 2004, California's governer Arnold Schwarzenegger unveiled plans to transform the state's 21 interstate freeways into hydrogen highways in six years, reducing greenhouse gas emissions from vehicles by 30 percent. Mini-networks clustered in the San Francisco Bay area–Sacramento region and the Los Angeles–San Diego region would eventually be linked together with a total of approximately 250 filling stations servicing some 20,000 vehicles. Public–private partnerships between the automobile industry, industrial gas corporations, energy corporations, government and academic institutions were expected to put the blueprint into action, and businesses were incentivized to build the hydrogen infrastructure with a 50 percent state subsidy.

While government intervention is often critical to create niches for new technologies that have to compete against "locked-in" technologies, there are a number of instances where strategic government interventions fail. The Hydrogen Highway has largely stalled in the face of California's cash-flow problems and the resurgence of electric vehicles among car manufacturers. Staying in California, Kemp, Rip and Schot (2001) describe efforts to promote wind-power as an example of when government subsidies can hurt rather than help a technology change. In this case, the subsidies aided the production of cheap, but poorly performing technology that was inefficient and thus ultimately unmarketable.

Given these problems, governments often shy away from trying to back winners, and focus instead on creating a general environment that encourages innovations, rather than being prescriptive about the form that these might take. This generates a "portfolio of experiments with different technological and social innovations" within sectors critical to sustainability, such as energy (Meadowcroft 2009: 325). As with many forms of governance, the overall target or end point is set, but the route there is left open (Geels *et al.* 2008).

Real-life experimentation prompts radical social and technical transformation, and places have sought to gain a competitive advantage by enhancing their capacity to innovate and become leaders in sustainability. This jockeying for position has produced some very high-profile projects. For example, Masdar City is an entirely new settlement of 50,000 residents in the desert 17 kilometers from Abu Dhabi in the United Arab Emirates. Designed by Foster and Partners to be a model of sustainable urban development, it promises to be a zero-carbon, zero-waste city powered entirely by renewable energy (Masdar City 2010). The Arabic word *masdar* means "source," and the project aims to generate ideas and knowledge to make Masdar City a global model for sustainable urban development. It is arguably the most ambitious attempt to use technology to tackle the issue of climate change. The city of Masdar itself will be used to test carbon-free technologies and lifestyles, and will be monitored by the newly established Masdar Institute, which lies at its core (Evans and Karvonen 2011). The ability to monitor, evaluate and eventually learn from innovations and experiments is seen as a critical element in allowing them to be rolled out at the regime level.

Society and transition

Despite the growing popularity of the transition concept in academic and policy-making circles, and the associated blossoming of experimental projects in the real world, the approach is not without its critics. One of the most basic observations is that innovation does not necessarily have to involve the creation of new technology. For example, Denmark did not invent the windmill, but has grown a successful clean technology industry around the world-leading wind turbine company Vestas by "re-innovating" an old technology. As Steward (2008) notes, a number of traditional technologies have the potential to contribute to sustainability, and they should not be overlooked in the relentless pursuit of the new.

In a similar vein, innovation does not necessarily have to be technological in character, but can be social or political, concerning new ways of living and social practices, or new ways of organizing governance functions. In fact, it seems quite reasonable that a transformation of society should require social innovation. Steward (2008) contrasts iconic big science projects from the past, like the Manhattan Project (which developed the atomic bomb) and the Apollo Program (which put a man on the moon), to public health and welfare reforms that led to the rolling out of comprehensive healthcare and benefits. The former may have generated radical technical innovations, but they did not transform everyday life or our economic system.

In contrast to top-down technological missions, the public health reforms of the nineteenth century and welfare reforms in the twentieth century were instigated in a piecemeal, bottom-up fashion by knowledge professionals, social movements and business entrepreneurs. It was only as they were taken up by political reformers (usually in the face of a perceived crisis) and incorporated into national policy that a wider transformation of the social system took place. Encouraging the adoption of sustainable lifestyles often involves simple changes rather than radical innovation, as discussed in Case study 7.3. Approaching sustainability as a high-level, high-profile technological problem may be to miss the point.

Case study 7.3

Cycling in the Netherlands

Around one third of all journeys made in the Netherlands are by bicycle, by far the highest proportion in Europe, comparing to less than 2 percent in the UK and less than 1 percent in Italy (Gilderbloom *et al.* 2009). The Netherlands is a highly developed country with a relatively affluent population who can afford to own cars if they desire. The weather is similarly wet and windy in the Netherlands and the UK, as is the level of bike theft. While the Netherlands is flatter than some parts of the UK, it suffers from severe headwinds as a result.

The factors that explain the huge discrepancy in use of the bicycle are related to the way in which it has been incorporated into the social and built environment. Perhaps most importantly, cyclists in the Netherlands are protected by strict legal liability, which means that in any collision with a motorized vehicle, the insurance of the motorist will be claimed against. This shifts the onus of responsibility for avoiding collisions decisively onto the motorist. The legal environment is complemented by favorable road design, which ensures a continuous network of separate bike lanes, often designed as dual carriageways, to maximize flow (Pucher 2007). The wider public realm is also designed to make bikes the easiest way to get around. *Woonerfs* are specially built neighborhoods in which schools, residences and workplaces are designed to be near enough to cycle between, while planners facilitate multi-modal transport, whereby bikes can easily be taken on trains and ridden straight onto ferries. Finally, the design of the ubiquitous utility bikes encourages a broad range of people to use them, being cheap and practical, with mudguards, built-in locks and so on.

Taken together, these factors have created a culture of cycling in the Netherlands, from school children who grow up cycling to school, to office workers and people on a night out. Because all sections of society engage in this form of transport it is normalized. Interestingly, rates of cycling in the Netherlands have not always been so high. In 1950 cycling levels were higher in the UK than they were in Amsterdam, with 15 percent of all trips being made by bike. As in the UK, rates of cycling plummeted in the Netherlands until the mid-1970s, when there was a massive reversal in transport policy, and the introduction of strict liability for motorists opened up a window of opportunity in the predominant automobile regime. There is no reason why the changes that created a cycling culture in the Netherlands could not be reproduced elsewhere.

Criticisms of technologically focused theories of transition also apply in part to the idea that the developing world can "leapfrog" the dirty technologies of the developed world and go straight from traditional to renewable technologies. While there is less lock-in to overcome, as technologies like the car have not saturated many less affluent societies, governments in developed countries have far fewer resources with which to create an economic and political environment that is conducive to innovation and the adoption of clean technologies (Perkins 2003). Unruh and Carillo-Hermosilla (2006) suggest that leapfrogging requires developed countries to lead the way by deploying low carbon technologies on a large scale, which can then be followed by developing nations. Unfortunately there are few good examples to follow—renewable energy technologies count for only 1 percent of global electricity production. Thus, they conclude, the prospect for leapfrogging is dim.

The idea that sustainable technology can simply be dropped in to developing countries conceals a whole gamut of social and political issues. This applies to the lock-in issue too. While societies may not have an actual dependence on dirty technologies, they can have an enormous symbolic dependence. Put simply, in many parts of the world factories and freeways represent economic progress. Changing cultural stereotypes and desires about what constitutes success and progress in a modern society (i.e. the landscape level) may represent the biggest challenge to creating a sustainable transition.

Talking about the social aspects of sustainable technologies, English sociologist Elizabeth Shove (2003: 9) suggests that preferences and needs are not stable and taken for granted, but, rather, are "immensely malleable." Education schemes designed to inform users of their environmental impacts play upon this very fact. For example, the Dawn Project, which aimed to raise public awareness about the adverse environmental impacts of increasing energy use in Thailand, developed educational materials on energy saving in daily life based on the concept of life-cycle assessment, and provided training among teachers and community leaders. The project engaged more than 300,000 students at primary and secondary schools, 23,400 teachers and 2,400 community leaders all over the country. Almost half of the schools demonstrated at least a 10 percent reduction in energy consumption. The field of socio-technical studies, discussed in Analytics of governance 7.1, addresses the way in which society and technology co-evolve, and has been used to understand the diffusion of sustainable technologies through society.

Analytics of governance 7.1

Socio-technical studies

Hoogma *et al.* (2002) argue that technological options, user demands, and institutions are created and shaped by the process of technology development. For example, the spread of the automobile in post-war America, which led to people traveling longer distances to work every day, is inextricably linked to the suburban sprawl that constitutes everyday life for many Americans today. Commuters do not sit in traffic for so many hours per day because they want to. Rather, dispersed settlement forms, the spatial separation of work and home, the post-war urban planning paradigm, home construction subsidies to World War II veterans, and vigorous political lobbying by the automobile industry all came together to explain the preference for today's living-room sized cars that are required to make long commutes tolerable (Brand 2005).

Co-evolution suggests that the relation between technology and society is perhaps more complex than the assumption that technology is invented in response to human needs, and then simply unleashed. Because the circumstances around us already exist when we come to the world, it is easy to forget that they have been designed and can therefore be redesigned (as the Dutch landscape, discussed in Case study 7.3, has been to accommodate bicycles). This is a potentially liberating realization, resonating with arguments that "the so-called environmental crisis demands not the inventing of solutions but the re-creation of *the things themselves*" (Evernden 1992: 123, emphasis in original). As Ralf Brand (2005: 13) states, "if we engage in serious dialogue between those who design, provide, organize and maintain these circumstances and those whose behavior and daily decisions are a reaction to them, we might discover their constructedness and hence their malleability." Increasingly, approaches to innovation and technology development are taking this message on board, involving users in the process of product design and application.

Confusion over what is being innovated leads to the more fundamental question of what is actually being "transitioned" in transition studies (Meadowcroft 2009: 326). For example, transition management assumes that it is possible to deal with something like the energy regime, if not entirely separately from related spheres of society like commodity chains, then at least analytically separate. Underplaying the social complexity of sustainable technologies runs the risk of neglecting the legal and political aspects of transition.

Moving from a sanitary model, whereby the state provides centralized services, to a sustainable model, whereby services are decentralized and managed by a broad range of actors, raises a series of social issues (Pincetl 2010). For example, even simple water-harvesting technologies such as butts, which are large plastic containers to capture rain flow off roofs for domestic or gardening use, have a series of social implications. The general shortages of water in some US states mean that homeowners are not allowed to prevent water running off their property into watercourses, making technologies to capture rainwater like butts illegal. Butts also have potential implications for public health. Obviously, such technology holds great appeal in hot climates, where gardens require more water but supply is under increasing pressure. But the heat also turns standing water into an incubator for diseases like Nile fever and malaria, raising questions about whose responsibility it is to maintain such technology and deal with anything that may go wrong. For the people who are supposed to adopt these technologies, these issues go beyond simply needing to know how to install them, completely recasting their relations to the state, water companies, neighbors, and even themselves (to whom they are now responsible for supplying water).

Attitudes to sustainable technologies are thus influenced by a range of cultural factors. As Yvonne Rydin (2010) has shown, homeowners often do not install energy insulation even though it will save them money if it conflicts with their daily practices. Subsidies to switch to sustainable technology need to fit with underlying habits and practices to produce changes in behavior (Owens and Driffil 2008). Case study 7.4, which explores the example of the adoption of the electric car, not in the twenty-first century but in the early twentieth, sheds light on the social, economic, and political aspects of technological transition.

Similar issues pertain to many sustainable technologies that change the way in which basic needs such as water and power are delivered. As for the water butts discussed above, renewable energy technologies move society from a centralized to a dispersed infrastructure. In the former, the state or a commercial body has sole responsibility for service delivery, and any problems that may arise are negotiated between the provider and the consumer. In the latter, a whole new set of actors suddenly become producers *and* consumers, complicating the simple division of responsibilities between public and private spheres. Because dispersed infrastructure produces dispersed responsibilities, seemingly simple technologies can have considerable social and legal implications.

Case study 7.4

Electric vehicles

As with many current sustainable technologies, the electric vehicle is nothing new. Management and entrepreneurship scholar David Kirsch's (2000) account of the Electric Vehicle Company, New York, which existed between 1897 and 1912, describes how steam, electricity and gas competed to become the leading automobile technology in the early twentieth century. The story yields a number of important lessons concerning the ways in which technologies diffuse. First, and perhaps most importantly, electricity was about a different service model of urban mobility, whereby people were not expected to own and maintain personal vehicles. Electric vehicles were intended to be leased or to operate as an army of taxicabs in built-up areas, while other forms of transport would be used for long journeys between cities. The eternal problem of the battery life of electric vehicles, which limits their range, disappears if it is possible to change the way in which people think about their mobility.

In fact, the incessant promises by makers and exponents of electric vehicles that a better battery was imminent may actually have put a brake on sales. As Kirsch asks, how many people would have bought an electric vehicle if there had not been the constant promise of a better battery just around the corner? No one wants to purchase technology that is instantly obsolete. Of course the irony is not only that batteries haven't improved all that much in the last hundred years, but that exactly the same claims are being made by makers of electric vehicles now.

A final lesson that emerges from the story of the automobile is that gas cars were themselves seen as a way to improve the environment when they were first introduced. In cities dominated by horses, manure was a serious problem. Gas cars by comparison were seen as clean and efficient. It is only the scale on which the gasoline-fueled automobile was subsequently adopted that made them a problem. At the start of the twentieth century, no one predicted either that any single technology would become dominant in the automobile sector, or that automobiles in general would become so widespread. The first lesson is that it is hard to fully predict or test the impacts of new technologies. The second is that there is only a relatively short window of opportunity to influence technology (in the case of the electric car in the early twentieth century, about 10 years when it competed on a par with gasoline-fueled vehicles before lock-in occurred). The difficulty with changing attitudes around personal mobility today is that the weight of history means that any electric vehicle has to be able to operate in a landscape shaped by the range and capabilities of the gasoline powered car.

Transition in practice

The most concerted and substantial effort to put the ideas of transition management into practice has been undertaken in the Netherlands. In 2005, the Ministry of Economic Affairs established the Energy Transition program to steer the Dutch transition to sustainable energy. Experimental projects were seen as central to innovation and real-life learning, and the program established what it called transition platforms, comprising networks of energy stakeholders who generated ideas for concrete projects. Projects were selected on the basis of costs and benefits, the likelihood of business investment, strength of demand, and chances of technical success. The first round of 70 projects began in 2005 with about €10 million of public money. This was stepped up to €15 million in 2006 and €20 million in 2007, supplemented with match-funding from private partners. Knowledge sharing between the projects was managed by an institution called the Competence Center for Transitions, established as a joint initiative between the Ministry of the Environment, academia, the Netherlands Organization for Applied Scientific Research and the SenterNovem Agency for sustainable development.

While projects were initially intended to facilitate open learning, they were gradually reframed as a means to create new business. Further, because the program remained bottom-up with a focus on experiments and projects, its influence on wider energy policy at the regime level (for example, regulations, energy markets, product standards, user behavior, and infrastructure renewal) has been limited so far. In the language of Geels *et al.* (2008), the Netherlands example has been dominated by the existing regime, with the result that few fundamental questions have been raised regarding the current regime, levels of energy consumption, dependence on other countries, social equity or who has and should have the ability to generate power. This is perhaps not entirely surprising, given that the key stakeholders were taken mostly from the established energy field (the program was chaired by the CEO of Shell in the Netherlands). In practical terms, windows of opportunity for wider diffusion were not created, while, abstractly, theorists have been left ruing the fact that transition management in this case has not lived up to the open, reflexive process it was intended to be (Kemp *et al.* 2007).

Emphasizing technical solutions also risks neglecting the needs and capabilities of specific places. In their study of the Clean Urban

Transport Europe program, which aimed to establish demonstration sites for green transport solutions in major European cities, Hodson and Marvin (2009) argue that demonstration projects are simply dropped in to urban areas rather than being developed with them. The corporate partnerships charged with innovating tend to focus on the ecological, technical and economic aspects of pilot projects at the expense of social issues, actually meeting with local resistance in some places. In these cases, the language of testing is indicative of attempts to simply field-test new technologies, rather than experimenting with genuinely new ideas and learning from them. Talking about London, they note, "a commitment to a socially more inclusive and highly participatory approach . . . coexists with a more top down experimental approach that sees London as the site for demonstration by a European partnership of multinational oil and automobile industries" (Hodson and Marvin 2007: 304).

Conclusions

Transition management focuses on steering society towards more sustainable futures, and is being taken up with great interest by governments and companies alike. Table 7.1 summarizes the key strengths and weaknesses associated with transition management. Perhaps its main limitation is that beyond the developed world context, it is unlikely that many countries have either the political or technical capacity to create the conditions necessary for technological change. Even in the developed world, levels of public commitment to both scientific research and the environment often fall short of those required by transition management. Pre-existing cultural norms can be critical; for example, the Japanese tradition of *mottainai*—"too precious to waste"—has allowed them to follow a remarkably similar trajectory to that of Germany and the Netherlands.

Technology is often seen as a panacea by the public and policy-makers alike. The lessons from the technological transitions literature indicate the importance of wider social expectations and political context in technological diffusion. Technologies imply new ways of living, but political vision is required in order to make sure that technologies enable the lives that we want. There is a danger in fetishizing technological innovation over social innovation, when many of the challenges of environmental governance are essentially political in their nature. This problem has hampered the success of transition

Table 7.1 ***Strengths and weaknesses of transition management***

Strengths	*Weaknesses*
Strong model of how steering can take place	Requires strong state commitment to technology and the environment
Recognizes key role of technology	Neglects the role of social and political innovation
Focus on innovation can drive change	Difficulty of establishing different visions within current system
Focuses on systemic transformation	Has tendency to overlook the needs of specific places, and can only operate within a strong political unit

management in practice, as it has struggled to generate genuine alternatives to the status quo.

Transition management ideally takes place within a strong political unit, like a state, region or city, capable of setting its own incentives. The German success story, whereby carbon emissions declined 22.3 percent between 1990 and 2007 (Boden *et al.* 2010) conceals the fact that much of Germany's dirty industry has simply been moved to China. Germans (and in fact most of the developed world) still consume products that generate high carbon emissions, it is just that they are made elsewhere and thus appear on the emissions balance sheets of other countries. Northern Europe has effectively outsourced its emissions. When national greenhouse gas emissions are measured in terms of what is consumed, rather than what is produced, a rather different picture emerges.

A lack of appreciation for the wider political context in which technology operates has produced a rather schizophrenic mentality among policy-makers, who shift between espousing grand technological solutions and individual behavior change. This policy stance does not address the urgency and radical nature of innovation demanded by the analyses of the IPCC and Stern reviews, which imply a need to re-introduce a social mission into the heart of innovation policy, and assist sustainable innovations to overcome lock-in and the vested interests of those who benefit from maintaining the status quo (Steward 2008: 5).

As an approach that deploys evolutionary economics to study systems, it is perhaps not entirely surprising that political agency takes a back seat, and even proponents of the transition approach recognize that the agent of evolutionary selection, which determines which niche experiments succeed and which fail, is viewed rather unproblematically (Geels 2002). Whether evolution is seen as a process of variation, selection and retention, or as an unfolding of combinations that create new trajectories, innovation is simply assumed to be a product of its wider economic environment. But the key selective pressures are anything but economic. For example, if left to the market alone, then the selective pressures of consumers may favor new technologies that are actually unsustainable (like 4×4s). In reality, transition management involves the state setting political goals, then backing specific technologies and creating a wider economic context in which they can be achieved. The technological transitions literature shows that transition is possible, even in democracies, but we need to better understand the governance arrangements that generate innovation and overcome lock-in.

Questions

- Who are the key actors in transition management, and how are their actions coordinated?
- Is it possible to steer a sustainable transition?

Key readings

- Geels, F. (2002) "Technological transitions as evolutionary reconfiguration processes: a multilevel perspective and a case study," *Research Policy*, 31: 1257–74.
- Kirsch, D. (2000) *The Electric Vehicle and the Burden of History*, London: Rutgers University Press.
- Meadowcroft, J. (2009) "What about the politics? Sustainable development, transition management, and long term energy transitions," *Policy Science*, 42: 323–34.

Links

- www.commutercars.com/. Website of a new super-thin electric commuter vehicle that explicitly addresses social habits and issues of lock-in.
- http://twitter.com/cleantechgroup. Twitter feed for news from a leading cleantech group.

8 Adaptive governance

Intended learning outcomes

At the end of this chapter you will be able to:

- **Understand resilience and the adaptive cycle.**
- **Appreciate how adaptive governance has been applied to social and ecological systems.**
- **Be aware of the strengths and weaknesses of resilience as a basis for environmental governance.**

Introduction

> Knowledge of the system we deal with is always incomplete. Surprise is inevitable. Not only is the science incomplete, the system itself is a moving target.
>
> (C. S. "Buzz" Holling 1993: 553)

Global financial meltdown, climate change and peak oil have highlighted the need for human society to adapt to future shocks, crises and disasters. Resilience has become rapidly ensconced within environmental policy and research agendas as a means to achieve this, and, in the process, make society more sustainable. The idea of resilience comes from a body of work in ecology which suggests that the persistence of ecosystems does not depend on their ability to remain stable in the face of change, but on their ability to shift between multiple states in the face of changing environmental conditions. Resilience is the "measure of the persistence of systems and of their ability to absorb change and disturbance and still maintain the same relationships between populations or state variables" (Holling 1973). The implications of this definition are that while relationships may persist after a shock, the system that maintains them may be different.

The concept of resilience is increasingly influential in the realm of environmental governance, holding considerable appeal to policy-makers who must adapt society to changes that cannot be predicted with any accuracy. Aware of the shortcomings of current environmental policy, advocates of resilience have cleverly positioned resilience and adaptive governance as a way to achieve sustainability, rather than as something separate, or incompatible with, sustainability. Resilience questions a number of cherished assumptions about how society should be governed. For example, systems that become highly evolved to one set of conditions may be very efficient, but they have little capacity to adapt to changes, making them less resilient. This chapter introduces the basic tenets of resilience and adaptive governance, shows how they can be applied to social–ecological systems, and considers its strengths and weaknesses as a mode of environmental governance.

Resilience and the adaptive cycle

The grandfather of resilience, American ecologist C. S. "Buzz" Holling (1973) defines resilience as a form of system persistence that does not depend upon traditional notions of stability, and makes a critical distinction between engineering resilience and ecological resilience:

Engineering resilience seeks to maximize the amount of disturbance that can be resisted by a system, and the speed with which it rebounds to its original state. This form of resilience maximizes efficiency, is controllable, and suits systems with low levels of uncertainty in which the potential levels of disturbance (or stress) are predictable. The design of suspension bridges to flex in high winds is an example of engineering resilience, where materials are chosen to withstand certain levels of stress without breaking.

Ecological resilience by contrast is persistent, adaptable, variable, and unpredictable. This form of resilience maximizes "the capacity of a system to absorb disturbance and reorganize while undergoing change so as to still retain essentially the same function, structure, identity, and feedbacks" (Walker *et al.* 2004: 5). A resilient system persists by adapting to disturbance and finding new states that allow it to continue performing its core functions. Ecological resilience measures the amount of disturbance required before a system becomes unable to continue performing its core functions. In terms of coastal management, the growing preference for managed retreat, whereby areas of land are sacrificed to the sea rather than protected through traditional flood defenses, is an example of ecological rather than engineering resilience (Vis *et al.* 2003).

Holling (1973) argues that persistent natural systems are not characterized by stability per se, but by resilience, and uses the example of occasional budworm outbreaks in the spruce-fir forests of eastern Canada to demonstrate this point. Budworm is exceptionally rare in these forests, being controlled by a range of natural predators, but is occasionally responsible for major outbreaks that destroy the mature fir trees, leaving spruce, white birch, and densely regenerating fir and spruce. Between outbreaks the firs tend to out-compete the spruce and birch, which suffer more from crowding, producing a forest dominated by firs. But given the combination of large numbers of fir trees and a succession of dry years, budworm populations escape the control of their predators and cause an outbreak. The outbreak ends when the budworm have destroyed so much of the fir population that they undermine their own food source and their population collapses, returning to its original background level.

As Holling points out, without the occasional outbreaks of budworm to control fir tree growth, the spruce and birch would be entirely lost. Periodic fluctuations in the form of budworm outbreaks are essential to maintaining the budworm, its predators and the diversity of trees in the forest. Seeming instability is necessary to maintain successive generations of species and the persistence of the system over some 300 years of recorded outbreaks. Holling argues that what looks like a highly unstable system over the short term, or a stable system with large parameters over the long term, is actually better described as a resilient system, which has the ability to adapt to disturbance and maintain the key relationships (or system functions) between populations. Freshwater lakes do something similar, flipping from a clear water state to a turbid state with algal blooms when nutrients from fertilizers or sewage make their way into them (a process known as eutrophication). While the clear water state provides more ecosystem services, the two states form part of a system that is remarkably resilient over time (Carpenter 2001).

Figure 8.1 depicts this model of dynamic stability through the example of a ball in a cup. The valleys represent different stability domains, the balls represent the system, and the arrows represent disturbance. As long as the ball remains within a single valley then it is displaying an engineering type of resilience, and its behavior can be described by the simple linear relationship between the size of the disturbance and the time it takes to return to rest at the bottom of the valley. Ecological

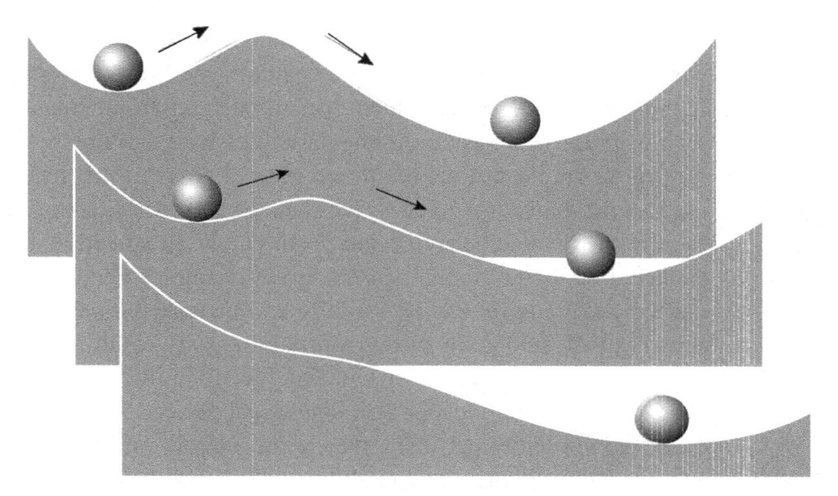

Figure 8.1 ***Ball and cup heuristic of system stability***
Source: adapted from Gunderson 2000: 427.

resilience is described by the width of the valleys within which the ball may come to rest, which represent a range of positions in which the system delivers largely the same functions.

The valleys are called "domains of attraction" because the system will be attracted to a different resting point once it enters a new domain. In the figure, the height of the ridge between the valleys indicates the amount of external disturbance that is required to force the system into another domain of attraction. The shape of the domain changes in response to shifting environmental and social conditions (for example, the valleys in Figure 8.1 might be representations of the same system as it changes over time). Drawing on the insights of chaos theory, resilience emphasizes the capacity of a system to occupy multiple stable states within a domain of attraction.

As Holling (1973: 15) argues, "although the equilibrium-centered view is analytically more tractable, it does not always provide a realistic understanding of the systems' behavior." In other words, ideas of stability and equilibrium appeal to us intuitively when we think of natural systems, but are to some extent mythical when we look past that which we can easily see (Deneven 1992). Paleoclimatologists, who study how the global climate has changed in the past, have shown that periods of stability and linear change are punctuated by tipping points (discussed in Key debate 8.1) when the climate shifts rapidly into new domains of attraction.

The application of engineering resilience to ecological systems leads to what resilience ecologists call the "pathology of resource management" (Gunderson *et al.* 1995), which occurs when environmental managers seek to preserve a system in one particular state and prevent it from moving to another. As Holling notes (1973: 15), "if this perspective is used as the exclusive guide to the management activities of man [*sic*], exactly the reverse behavior and result can be produced than is expected." Taking biodiversity, he gives examples where the traditional link between environmental stability and increased diversity fails, showing that, in certain systems, instability may result in a higher diversity of species across space, and "hence in increased resilience" (ibid.: 19). Holling's paper also distinguishes between the use of quantitative (numerical) data, which are suitable for measuring engineering resilience, and qualitative (descriptive) data, which are more suitable for measuring ecological resilience. Numbers work well for linear systems characterized by engineering resilience, but have a limited ability to detect or describe the kind of system changes that occur at tipping points. As a result, "our traditional view of natural systems, therefore, might well be less a meaningful reality than a perceptual convenience" (ibid.: 1).

Resilience views ecosystems as complex adaptive systems, whose resilience is defined by their ability to adapt to change (Levin 1998). Figure 8.2 shows the adaptive cycle, characterized by four phases (r, K, Ω, and α), which describes how a system adapts to external shocks (Redman and Kinzig 2003). In the r-phase the system grows under conditions of stability, accruing more capital or resources (represented on the y-axis), and becoming increasingly interconnected (represented on the x-axis). In the K-phase the system has become closely adapted to its environment and most of its capital is stored up in the system, perhaps in the form of biomass. The Ω-phase constitutes the release of this capital, prompted by a major external or internal shock to the system. Returning to our examples above, this might be the succession of warm summers that causes an explosion of budworm, or the flux of nutrients into a lake. The Ω-phase represents the rapid breakdown of the system as both capital and complexity decrease. Finally, the α-phase signals reorganization and growth, as the system adapts to its new environmental conditions and reforms. As the system deals with periodic shocks the cycle repeats itself, but always following slightly different patterns of regrowth within its overall domain of attraction.

Key debate 8.1

Tipping points

Emissions from human activity have caused the atmospheric concentration of greenhouse gases to rise from 280 ppm of carbon dioxide before the Industrial Revolution to approximately 390 ppm now. Even if we stopped all emissions tomorrow, the world would continue to warm by at least another half degree by 2050. Further, this warming would have knock-on effects on hydrological cycles, ecosystems, sediment cycling, and the societies that depend on them. The oceans could be expected to absorb about 90 percent of the excess carbon dioxide in the 1,000 years that it takes for surface and deep waters to completely turn over. The geologic cycle would eventually mop up the rest, as carbonic acid formed by atmospheric carbon dioxide mixing with rain weathers rocks like feldspar and quartz to form carbonates that are then washed out to sea. This final stage would take some 100,000 years to complete (Weisman 2007). Like the proverbial stone in a millpond, the ripples caused by our actions will be felt widely for a long time.

Unfortunately, climate change is not a mathematical problem that has a perfect solution, but a highly complex, dynamic system (Auld *et al.* 2007). The global atmospheric–oceanic system cannot be depended on simply to return benignly to its original equilibrium point, even if we eventually reduce atmospheric carbon to its original levels. Scientists have identified a number of so-called tipping points in the system, at which rapid and often irreversible changes occur (Lenton *et al.* 2008). Classic examples in climate science are related to the melting of the ice caps. For example, being white, ice has a high albedo, reflecting solar radiation away from the Earth's surface. As the ice caps melt, the Earth's surface will darken, absorbing more and more of the Sun's heat, causing ice to melt faster and faster. This runaway effect, known as positive feedback, can cause a system to flip rapidly into a completely different state—in this case a much warmer world with no ice and sea levels approximately 80 meters higher than they are now (United States Geological Survey 2007).

Returning to Figure 8.1, a tipping point occurs when the ball falls over the edge of a ridge into another valley and comes to rest in an entirely different state. Tipping points represent an extremely dangerous aspect of climate change, as they are rapid, sweeping, irreversible (on human timescales at least), and the point at which they will occur is largely unknown. Because tipping points represent non-linear change they are hard to predict and hard to prepare for. As Pahl-Wostl (2007: 51) states in relation to water supply, "improving our understanding of the likelihood of extreme events based on experience derived from historical records does not tell us much about the likelihood of extremes in the future given the uncertainties caused by climate change." A major rationale for limiting global warming to 2°C is that the likelihood of triggering major tipping points increases significantly once this threshold is passed.

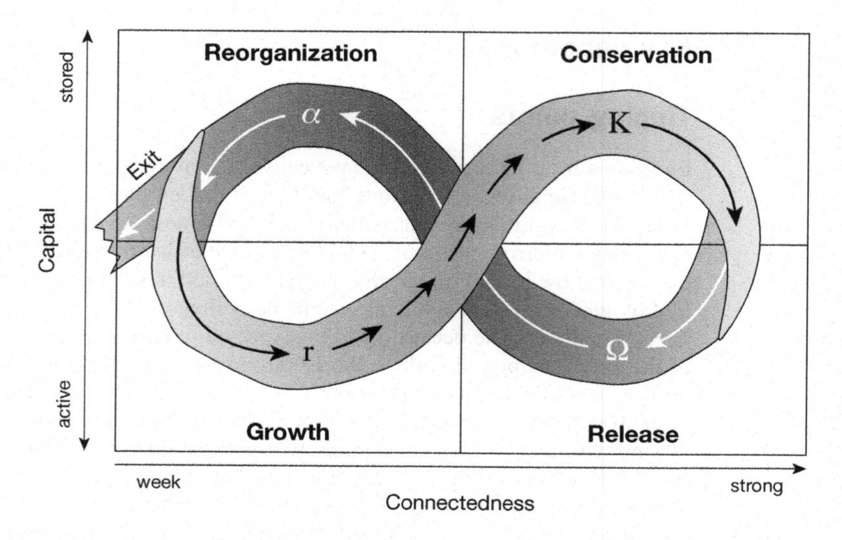

Figure 8.2 ***The adaptive cycle***
Source: adapted from Gunderson and Holling 2002.

Social–ecological systems

The insights of ecological resilience have been extended to explore environmental questions through the concept of the social–ecological system (SES). Redman *et al.* (2004: 163) define a social–ecological system as a "coherent system of biophysical and social factors that regularly interact in a resilient, sustained manner . . . with continuous adaptation." Berkes *et al.* (2001) highlight the linkages between social and ecological systems, arguing that previous studies either ignore or black-box one or the other of these components. They argue that while the traditional concept of the ecosystem is deficient to describe the complexity of social processes that are characterized by human intentionality, the broader concept of the system can be extended to cover them. While resilience has slightly different meanings in social and ecological contexts (Adger 2000), the social–ecological system approach holds that both social and ecological systems display resilience, are complex, and are linked through feedback mechanisms. Figure 8.3 depicts a social–ecological system as two nested systems linked by ecosystem feedback and management practices.

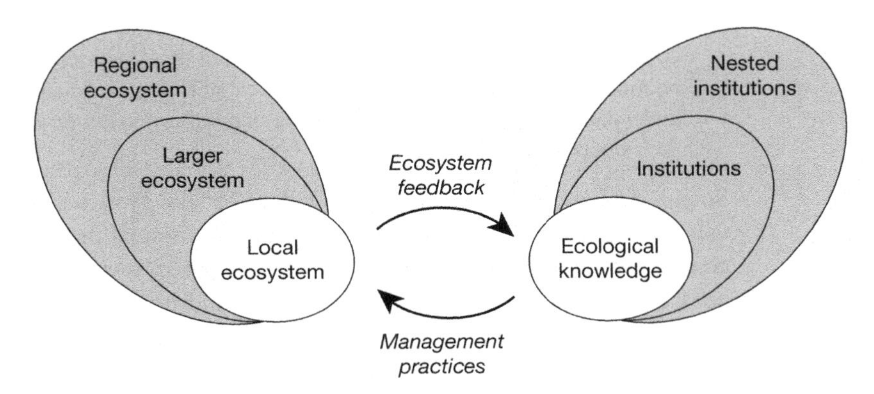

Figure 8.3 ***A conceptual framework for analyzing social–ecological systems***
Source: adapted from Berkes *et al.* 2003.

Berkes *et al.* (2001) identify four dimensions of a social–ecological system: ecosystem, local knowledge, people and technology and property rights institutions. As Figure 8.3 shows, knowledge about the local ecosystem is essential to the functioning of a social–ecological system, and must be captured by institutions that can translate it into management practices. Similarly, Scheffer *et al.* (2002) identify four key ingredients for resilient sustainable human–nature interactions: clear understanding of ecosystem dynamics; clear understanding of social dynamics; avoidance of stakeholder bias, and social networks that bridge horizontal and vertical gaps. Focusing on ecological and social dynamics allows institutions to adapt to changing conditions, and understand the impacts of different decisions on social–ecological systems. Scheffer's final two ingredients (stakeholder bias and social networks) clearly resonate with the tenets of network governance, advocating a consensual model of decision-making based upon a dense network of social relations between stakeholders.

Drawing heavily upon common pool resource theory (discussed in Chapter 3), Berkes *et al.* (2003) suggest that social systems comprise governance, property rights, and access to resources (including knowledge, views and ethics relating to this access), and that community resilience requires diversity, knowledge, and self-organization. Berkes (2004: 628) states:

> Communities . . . do not act as simple, isolated agents. Rather they are embedded in larger systems, and they respond to pressures and incentives. It may be more useful to re-think community based

> conservation as shorthand for environmental governance and conservation that starts from the ground up but deals with cross-scale relations. To ground conservation effort, we need a more nuanced understanding of the nature of people, communities, institutions and their interrelations at various levels.

Social–ecological systems link physical and social systems together using multi-scalar feedback loops. A staple example is that of land-use change, which affects ecological pattern and process, which then feeds back into the social system to drive further land-use change. So, for example, the development of urban green spaces may decrease biodiversity and increase flooding problems. In a resilient social–ecological system, knowledge of these ecosystem changes would be captured effectively by institutions and fed back into decision-making to alter the management practices to increase the area of green space within the city, or set critical levels of green space that must be maintained across the city.

Adaptive governance

Adaptive governance focuses on increasing the resilience of social–ecological systems by enhancing their capacity to adapt. The resilience of social–ecological systems is related to the magnitude of the shock that the system can absorb and remain within a given domain of attraction, the degree to which the system is capable of self-organization, and the degree to which the system can build the capacity to learn and transform itself (Folke *et al.* 2002: 438). The question for governance is what features of a social–ecological system create the largest set of stable states within a domain of attraction, rather than what features enable it to resist and bounce back from disturbance.

Strengthening the ability of social–ecological systems to provide desirable services requires at least three levels of understanding: the dynamics of the ecosystem, the management of the ecosystem, including generation of knowledge and its use, and the institutional dynamics including governance and learning (Elmqvist 2008). Case study 8.1 discusses how these factors differ in practice.

In recognizing the importance of innovation and learning in coping with change, adaptive governance advocates an experimental approach to governing. Holling (2004) himself calls for the creation of conditions

Case study 8.1

Adaptive governance in the Florida Everglades and Grand Canyon

Folke *et al.* (2002) contrast case studies from the Florida Everglades and the Grand Canyon to highlight the key requirements of adaptive governance. Both are complex social–ecological systems that have experienced undesirable degradation of their ecosystem services, but they vary dramatically in terms of their institutional make-up. In the Everglades the governance structure is dominated by the interests of environmentalists and the agriculture lobby, who have historically conflicted over the need to conserve habitat at the expense of agricultural productivity. These tensions prevent the two groups from working together, with the result that there are few institutional feedbacks between the ecological system and the social system, and the social–ecological system is unable to innovate and adapt (the α-phase of reorganization and growth depicted in Figure 8.2).

By contrast, in the case of the Grand Canyon, stakeholders have formed an adaptive management workgroup, which uses planned management interventions and monitoring to learn about changes occurring in the ecosystem, and the best ways in which to subsequently manage them. This governance arrangement allows for institutional learning to take place, and successful reorganization and adaptation. Such an approach to institutional learning is becoming more common as NGOs, scientists and communities collaborate to manage ecosystems. Social network analysis, discussed in Chapter 5, is used extensively in studies of adaptive governance to understand the relationships between stakeholders in a social–ecological system. Drawing on the techniques discussed in Analytics of governance 5.1, a number of characteristics concerning the resilience of a social–ecological system can be inferred from its levels of connectivity and centrality (Janssen *et al.* 2006).

that promote many low-cost innovative experiments in governance. Similarly, Berkes *et al.* (2003: 433–34) state,

> adaptive management therefore views policies as hypotheses—in that most policies are really questions masquerading as answers. Since policies are questions, then management actions become "treatments" in the experimental sense. The process of adaptive management includes highlighting uncertainties, developing and evaluating hypotheses around a set of desired outcomes, and structuring actions to evaluate or "test" these ideas.

Adaptive governance involves creating institutions that have the capacity to experiment with different solutions and learn from them in order to adapt and transform.

Gunderson (1999) identifies three barriers to this kind of adaptive management: inflexible social systems, ecological systems that lack resilience, and the technical challenges associated with designing experiments. The work of Pahl-Wostl (2007) directly addresses how institutions must change in order to meet these challenges, exploring the implications of social learning and adaptive governance for water management. For her, adaptive governance is a proactive management style that attempts to enhance the capacity to change the structure of a system, not just respond to change. In relation to water, this might mean the ability to change the types of crops that are grown in an area, adapt the life-styles of water consumers or shift the allocation of water quotas between different users, rather than simply building more reservoirs.

An adaptive institution must be able to gather new information, process it and transform in response to it. Table 8.1 contrasts the traditional command-and-control model of governing to adaptive governance. Adaptive governance requires institutions to work in very different ways, engaging a broad set of stakeholders and operating across different sectors and scales. Traditional approaches to information are highly proprietorial, with data often being guarded closely by those institutions that own it. In contrast, adaptive governance operates on a model of open sharing, whereby different institutions pool their information in order to fill gaps in knowledge and facilitate integration. While traditional resource management monitors a narrow range of environmental variables, adaptive governance looks at a far broader set of variables, which may include social factors like the quality of communication in social networks or the appropriateness of a chosen institutional setting to facilitating experimentation and learning, as well as ecological variables.

Echoing the discussion of transition in the last chapter, adaptive governance implies a more decentralized infrastructure, with diverse elements. Similarly, adaptive governance seeks to use a portfolio of funding approaches, like public–private partnerships and market mechanisms referred to in Chapter 6. Shifting to an adaptive mode of resource governance involves wholesale changes to the ways in which institutions operate, and there are a number of factors that can cause reluctance, including the high costs of information gathering and

Table 8.1 ***Command-and-control versus adaptive governance***

	Command-and-control	*Adaptive*
Management paradigm	Prediction and control based on an engineering approach	Learning and self-organization based on a complex systems approach
Governance	Centralized, hierarchical, narrow stakeholder participation	Polycentric, horizontal, networked stakeholder participation
Sectoral integration	Sectors separately analyzed resulting in policy conflicts and emergent chronic problems	Cross-sectoral analysis identifies emergent problems and integrates policy implementation
Scale of analysis and operation	Transboundary problems emerge when river sub-basins are the exclusive scale of analysis and management	Transboundary issues addressed by multiple scales of analysis and management
Information management	Understanding fragmented by gaps and lack of integration of information sources that are proprietary	Comprehensive understanding achieved by open, shared information sources that fill gaps and facilitate integration
Environmental factors	Quantifiable variables that can be measured easily	Qualitative and quantitative indicators of whole ecosystem states and ecosystem services
Infrastructure	Massive, centralized infrastructure, single sources of design and power delivery	Appropriate scale, decentralized, diverse sources of design and power delivery
Finances and risk	Financial resources concentrated in structural protection (sunk costs)	Financial resources diversified using a broad set of private and public financial instruments

Source: adapted from Pahl-Wostl 2007: 55.

monitoring, the unfamiliarity of gathering new types of information, resistance from managers who may fear increased transparency and loss of control, and the political risks of failure (Pahl-Wostl 2007). While these are very real problems, the tenets of adaptive resource management are increasingly influential.

The politics of resilience and adaptation

Resilience has spread rapidly within the environmental policy arena, and, in promising a way to adapt to changes in underlying environmental conditions and cope with extreme and uncertain events, its appeal to policy-makers is fairly obvious (Evans 2011). But applying ecological theory to the social world is not without its problems. In terms of the ecological basis from which the concept of resilience emerged, Holling claims to be as surprised as anyone at the enthusiasm with which his ideas have been taken up by other disciplines. Holling's ideas were originally received with a degree of ambivalence within his home discipline of ecology. Here is not the place to recapitulate this debate, but it is worth noting that the insights of resilience have not been accepted as having relevance to all ecological systems. Cynically, the take-up of a scientific term like resilience in other social and political disciplines may reflect a desire to lend scientific credibility to modes of environmental governance that are essentially political. In the case of resilience, there can be little doubt that its most vocal exponents have mobilized the authority of ecological science to promote it as a policy discourse.

On the other hand, resilience thinking really does offer a handle on how to adapt under conditions of uncertain change. The adaptive paradigm conceptualizes social and ecological systems in a holistic way, while acknowledging their inherent unpredictability. Under conditions of uncertainty, systems cannot be "knowable," only "changeable," as the observer forms part of the system being studied. The language of generally applicable knowledge is replaced by a search for generally valid guiding principles, meta-principles and frameworks for how experimentation should progress to produce sustainability and resilience. Such an emphasis on procedure echoes the approach of governance more generally, but its acceptance of change has opened adaptive governance (and the concept of resilience upon which it is based) up to political critique, discussed in Analytics of governance 8.1.

Analytics of governance 8.1

The political economy of resilience

Resilience implicitly accepts many of the principles of free market economics. By naturalizing crises as inevitable, the adaptive cycle reproduces a fairly right-wing and insular discourse of individuals (or communities or cities) fending for themselves in the face of them (Evans *et al.* 2009, Walker 2009). As Berkes *et al.* (2003: 3) state in relation to the social ecological systems approach, "we consider change and the impact of change as universal." Resilience not only normalizes crisis and change, but, in establishing the logic of adaptive learning as coterminous with capitalist development and ongoing change, privileges those with the sufficient economic and intellectual capital who are best able to experiment and learn. What Caldeira and Holston (2005: 411) call the "complex relationship . . . between democratic and neoliberal planning" rests upon a shared acceptance of contingency and uncertainty, and resilience certainly buttresses this relation. Of course there is an irony here, in that on its surface resilience views capitalism and its drive for efficiency as detrimental to adaptive capacity.

This bias can be partly explained by the institutional context in which resilience has been established. The Resilience Alliance, based in the Beijer Institute in Stockholm, has seamlessly married its goal to establish an international network of influential environmental scientists working on resilience to an aggressive campaign promoting resilience as a policy priority. The Institute was established in 1977 and was reorganized in 1991 with a focus on ecological economics. As Walker (2009) has argued, the ecologists at the Resilience Institute have worked closely with neoliberal economists to develop their ideas concerning social ecological systems and resilience.

Applying ecological ideas of change and adaptation directly to social systems is appealing, as it promises to uncover a scientific basis for managing social processes and making decisions. But the evidence base that underpins the insights of ecological resilience was not developed in a social context and should not be unquestioningly applied to society. What we find instead in the resilience writings on social–ecological systems are fairly general allusions to the collapse of the Soviet Union, or the spread of telephony and cycles of commercial innovation and obsolescence, to demonstrate the need for adaptability (Perrings 1998).

Just as sustainability doesn't tell us what we should sustain, so resilience doesn't tell us what should be made resilient. It is important not to fetishize resilience as an end in itself—a polluting chemical factory might be organized in such a way as to be highly resilient, but that doesn't make it desirable. Cockroaches are extremely resilient organisms. To be fair, resilience thinkers are acutely aware of this problem. Berkes *et al.* (2001: 131) make a big play of needing to ask what direction society wishes to go in at the outset of any form of resilience planning, while Holling suggests that the adaptive cycle requires some form of political engagement in order to determine what kinds of social–ecological systems we want to create. Similarly, Gunderson and Holling (2002: 32) state that "the purpose of theories . . . is not to explain what is; it is to give sense to what might be."

Finally, resilience threatens to depoliticize highly political aspects of social transformation concerning how we wish to live, focusing instead on the establishment of technical feedback loops between environmental change and political decision-making, and an experimental mode of governance that remains dominated by experts (Evans 2011). Civil society and local communities have a critical role to play in adaptive governance, as they are the holders of the ecological knowledge that links social and ecological systems together, but, as with criticisms of governance more generally, this bottom-up perspective may not be capable of delivering the wide-ranging social transition that it is argued is required (Andersson 2007).

Conclusions

Given that the edifice of traditional ecology is based upon the concept of equilibrium, whereby healthy ecosystems remain in balance with their surrounding environment, the suggestion that stability may not be a useful basis for environmental governance is fairly revolutionary. Attempts to "preserve" natural systems in one particular state might actually be working against ecological processes, weakening rather than protecting the system. Resilience, and the concept of adaptive learning to which it is closely linked, thus represents an approach to environmental governance that embraces uncertainty. This has given it particular traction in the environmental field, and an increasing prominence in policy.

Table 8.2 summarizes the strengths and weaknesses associated with adaptive governance. The main appeal of adaptive governance is that it

Table 8.2 ***Strengths and weaknesses of adaptive governance***

Strengths	*Weaknesses*
Holistic understanding of environmental issues	Difficulty of identifying discrete SESs for certain environmental issues
Ability to adapt to change	Passively accepts change
Nested institutions	Difficulty of scaling up from local specificity
Emphasis on experimentation and learning	Practical difficulties of experimenting in real world
Links institutions to ecosystem knowledge	Reduces decision-making to technical feedback process

understands society and the environment as part of a single system, and foregrounds their ability to adapt to change. But while it is relatively easy to identify social–ecological systems for more localized resource use issues, this approach is less well suited to the conceptualization of a global problem like climate change. As discussed in the previous section, the foregrounding of change as a constant condition has also led to criticisms that adaptive governance is too passive, simply accepting changes that may be undesirable and avoidable. Within this understanding, knowledge about processes at one level cannot simply be scaled up or aggregated to understand process at larger scales.

Adaptive governance produces innovative responses to change through experimentation and learning, but raises a series of practical questions about how to persuade risk-averse managers and decision-makers to embrace a more exploratory way of operating. The key challenge is to design institutions and decision-making procedures that strengthen the feedbacks between socio-economic activity and ecosystems in order to adapt to environmental change. There are costs involved with experimenting and learning—doing lots of different things will always cost more than simply rolling out a single response, while monitoring and evaluation are essential in order to learn. Similarly, the focus on monitoring feedbacks between social and ecological systems runs the risk of reducing political questions concerning the future direction in which society should travel to a technocratic public participation process.

That said, it seems premature to simply write off resilience as a flawed ecological model that has little bearing on social systems. The necessity to make decisions in the face of a changing environmental and political context makes it of unquestionable value to a specific set of governance challenges. Our current world of economic growth based on increasing global integration has proven very resilient in preventing the escalation of wars into world wars, but is increasingly fragile in the face of economic and environmental shocks. The resilience "of what, to what" is a critical question, and one to which the next chapter turns.

Questions

- Adaptive governance is only viable at small scales. Discuss.
- Is adaptive governance simply a form of network governance which includes ecological variables?

Key readings

- Armitage, D. (2010) *Adaptive Capacity and Environmental Governance*, Berlin: Springer.
- Folke, C., Carpenter, S., Elmqvist, T., Gunderson, L., Holling, C. and Walker, B. (2002) "Resilience and sustainable development: building adaptive capacity in a world of transformations," *Ambio*, 31: 437–40.
- Holling, C. (1973) "Resilience and stability of ecological systems," *Annual Review of Ecology and Systematics*, 4: 1–24.

Links

- www.youtube.com/profile?user=stockholmresilience#p/u/23/tXLMeL5nVQk. Brian Walker from Australia's national science agency, the Commonwealth Scientific and Industrial Research Organization, explains resilience in a video produced by the Stockholm Resilience Institute.

9 Participation and politics

Intended learning outcomes

At the end of this chapter you will be able to:

- **Understand the concept of risk and the precautionary principle.**
- **Articulate the basic premises and practices of participation in environmental decision-making.**
- **Express the post-political critique of participation and governance.**
- **Appreciate how actions outside of mainstream political channels influence environmental governance.**

Introduction

> Democracy is not a spectator sport.
>
> (Lotte Scharfman, 1928–70)

Involving the public in environmental governance makes intuitive sense. Many environmental decisions directly affect the public, from siting wind-turbines at the local level to taxing fossil fuels at the national level. Further, local communities have deep knowledge about and emotional attachment to the places in which they live and work, making them indispensible partners in the delivery of sustainable development.

While all governance is about engaging wider groups in governing, it is essentially about procedures, or how things should be done, rather than *what* should be done. For example, a powerful network is not necessarily a good thing in its own right—it depends to what end it is being put. As Banerjee (2008) points out, the terrorist group al Qaeda is a hugely powerful network with considerable social capital and highly effective procedures. Media guru Clay Shirky tells the story of the Sudanese government, which organized an anti-government protest via

Facebook and then arrested everyone who showed up (Burkeman 2011). There is nothing necessarily progressive or sustainable about networks per se.

Or, for that matter, markets. Adam Smith, the Scottish political economist and supposed grandfather of neoliberalism, preceded his oft cited opus *The Wealth of Nations* with his lesser known book, *The Moral Economy*, in which he argued that markets would only function correctly in a society that had a strong and shared set of basic values. Networks and markets can be used to steer society, but they do not tell us in which direction we should steer. Similarly both transition management and adaptive governance require participation in order to set their agendas for change (Gunderson and Holling 2002, Walker *et al.* 2002). In this sense, participation cuts across the other four modes, providing the substantive vision of *where* to steer society that informs the purely procedural concerns of *how* to steer society. Social and political values are an essential element of meta-governance, which can get overlooked in the emphasis on institutions and rules. To adapt an old adage, environmental governance without people is like the bus that always runs on time because it doesn't stop.

This chapter begins by discussing the risk society thesis, which argues that many of the environmental problems that we face today have actually been created by modern progress. Such risks suggest that the participation of non-expert groups, like the public, in environmental governance is necessary to steer society in desirable directions. The chapter then explores how governors enroll different publics in environmental governance through formal public participation. While there are a number of good reasons to involve communities in governance, formal modes of participation have received considerable criticism for failing to meaningfully influence decisions. The final part of the chapter considers how actions taking place outside of formal political channels influence environmental governance.

Risk

One of the deepest analyses of how modern progress affects society has been developed by the German sociologist Ulrich Beck. He argues that risk not only plagues modern society, but defines it. For Beck, industrial society was concerned with distributing the fruits of its labor, like services and products, but, at some point in the latter half of the twentieth century, the negative side-effects of progress and technology began to outweigh the positives. Decisions began to be taken in the interests of technological and economic gain, which accepted hazards as "simply the dark side of progress" (Beck 1992a: 8). Rather than hazards being a stroke of fate, attributable to chance or the will of some god, they were a direct result of political and economic decision-making. The globalization of corporate influence, coupled with the increasing capabilities of science and the dominance of technical expertise in decision-making, led to a proliferation of potential accidents waiting to happen. Toxins in foodstuffs, the threat of nuclear war or disaster like Chernobyl, and global warming are among many risks that have been unwittingly produced as side-effects of economic progress. Within what Beck terms the "risk society," the key question for those governing society becomes one of who should live with the "bads," rather than of how to distribute the "goods."

Modern risks display three characteristics that mark them apart from pre-industrial hazards. First, risks are geographically delocalized, so that the negative consequences of decisions are often felt far away. Second, the potential consequences of disasters are largely hypothetical. Finally, it is often impossible to compensate those affected in the case of a disaster. Climate change offers a good example of a risk that is hard to perceive, hard to quantify, and even harder to compensate for. The effects of climate change on atmospheric CO_2 levels cannot be directly sensed by humans, and ensuing disasters wrought by temperature changes will not necessarily affect those places that have been responsible for causing them, and may span many generations. The risks associated with global climate change are largely impossible to pin down, with estimates ranging from very little impact to the almost complete destruction of life on Earth.

When these aspects of risk are taken together, it becomes clear that the traditional mechanisms through which society mitigates against them, like insurance and compensation that are based on the probabilities of something bad happening, cannot be applied in any satisfactory manner. As Jean-Pierre Dupuy (2007) asks, who can really say whether a 0.6 percent probability of a Chernobyl-style reactor meltdown in a 50-year period is acceptable, given the dire consequences for those who live and work near nuclear facilities? In Beck's parlance, modern risks have become uninsurable.

In response, governors have developed a set of reflexive mechanisms to mitigate risks, like the precautionary principle. Writing 900 years ago, Christian scholar Saint Thomas Aquinas observed "it is better for a blind horse that it is slow." This, in a nutshell, is the ethic behind the precautionary principle, which urges society to proceed carefully in the face of unknown risks. The modern precautionary principle derives from the German concept of *Vorsorgeprinzip*, which balances economic gains against the achievable maintenance and improvement of environmental quality. Practically, precaution means taking thoughtful action in advance of scientific proof of cause and effect, leaving ecological space for ignorance, and taking care in management—particularly through the involvement of the public. The precautionary principle underpins almost all multilateral environmental agreements. For example, the Rio Declaration states that,

> (i)n order to protect the environment, the *precautionary approach* shall be widely applied by States according to their capabilities. Where there are threats of serious of irreversible damage, lack of full scientific certainty shall not be used as a reason for postponing cost-effective measures to prevent environmental degradation.
>
> (emphasis added)

More specifically, in relation to ozone depletion, the Montreal Protocol states that, "(a)lthough aware that measures should be based on relevant scientific knowledge, the Parties are determined to protect the ozone layer by taking *precautionary measures* to control equitably total global emissions of substances that deplete it" (emphasis added).

Not everyone agrees with Beck. While following a very similar line of argument, English sociologist Anthony Giddens (2002) is more optimistic in his prescriptions. While risk must be considered carefully, he holds that the capacity to take risks is an essential part of any dynamic and innovative society and should not be discarded. The American political scientist Aaron Wildavsky (Douglas and Wildavsky 1982) argues that precautionary approaches to new technologies are irrational, because they make it impossible to gather the very knowledge that is required to know what is safe and what is not. (This argument brings to mind the Buddhist parable of happiness, in which a man goes to his guru and enquires as to the secret of happiness. The guru replies, "good judgment," to which the man cries, "but how am I to get good judgment?" "Bad judgment" replies the guru.)

Wildavsky also argues that the negative consequences of environmental disasters are less severe than they are often perceived to be, and are far outweighed by the improvements in living standards that new technologies bring. For him, emphasizing the risks of new technology is both unhelpful, as the benefits tend to outweigh the costs, and ironic, as it is only in affluent societies that have benefited from technological advances that people can afford to worry about environmental risks at all. Like the experiments advocated by transition management and adaptive governance, he advocates a system of trial and error, whereby lots of alternatives are tested in order to enhance society's ability to adapt to the unexpected, rather than trying to prevent accidents from ever happening. That said, Wildavsky wrote for the most part before climate change had become a recognized threat. The trials that he had in mind were intended to be small-scale, not conducted on the entire planetary system.

Whether Beck's risk society thesis is accepted in full or not, it reflects a number of broader political trends. Trust in decision-makers has waned, scientists are now seen to have been wrong in the past, technology has produced unwanted side-effects, and the severity of many risks remains unknown while the long-term effects of others are only now coming to light. Risks are also subjective. What may constitute a risk to one person will not be to another, and the actual probability of an event occurring does not determine the amount of importance that is attached to it: a trend that the media tends to exacerbate. A key prescriptive element of Beck's (2007) thesis addresses the need "to debate, prevent and learn to satisfactorily manage risk, with the aim of facing up to the induced political hysteria and a perception of fear, often spread through common practices used by the mass media." For Beck, our current democratic system is no longer fit for purpose, developed as it was to distribute national "goods" rather than global "bads." The implications of risk are that new institutions are required that are capable of engaging the public effectively in decision-making in order to determine which risks are acceptable, and which are not.

The rationale for participation

Given the inherent uncertainty that plagues environmental problems, "there is a widespread appreciation that governments cannot legitimately keep up the idea that decisions can only be made on the appropriate knowledge available" (Hajer and Wagenaar 2003: 10). Stirling (1998: 103) emphasizes the need for public participation in order to make decision-makers more accountable—

> no matter how much information is available, and no matter how much consultation and consideration are involved, no purely analytical

> procedure can fulfill the role of a democratic political process . . . there can be no uniquely "rational" way to resolve contradictory perspectives or conflicts of interest.

As discussed in Chapter 1, environmental issues pose wicked policy problems, which require decision-makers to choose between imperfect solutions.

There are compelling ethical, practical and substantive reasons to involve communities and the broader public in environmental governance. Ethically speaking, citizens should have the ability to contribute to decisions that will affect them. Participation in decision-making extends the logic of democracy itself, which is predicated upon involving people in choosing their own government. Rather than simply presenting environmental problems as an external threat, communities and the public have a right to create the kinds of places and societies in which they want to live (Irwin 1995). Practically, involving people in making decisions is the most effective way to secure legitimacy for the decisions that are taken. Community engagement and public participation reduce conflict between different interests around contentious issues. As Walker *et al.* (2002: 14) note, "expert solutions may maximize something, but they rarely maximize legitimacy."

Finally, involving communities and the public in governance makes instrumental sense, by improving the quality of decisions. Only recognizing expert knowledge as a valid basis for decision-making excludes the knowledge and experience of people who live and work in ecosystems (Taylor and Buttel 1992). In contrast to expert knowledges, so-called lay knowledges are increasingly valuable to decision-makers in the wider context of the scientific uncertainty surrounding environmental questions. As Fischer (2000: 222) argues, "participation is not only seen as a normative requirement for a democratic society but serves increasingly as a counter to the uncertainties of science." It is precisely these qualities of lay knowledge that form the basis for Ostrom's common pool resource management, examined in Chapter 3.

Participation forms a central strand running through environmental policy. The Brundtland Report of 1987, *Our Common Future*, established sustainable development as its guiding principle, which emphasizes the inclusion of all elements of society in environmental decision-making. This commitment was brought to fruition in the Rio Declaration on Environment and Development and Agenda 21, signed at the Earth Summit in 1992, which states that "environmental issues are best handled with the participation of all concerned citizens, at the relevant level . . . each individual shall have appropriate access to information concerning the environment . . . and the opportunity to participate in the decision-making process" (United Nations 1992). The Rio tagline "Think global, act local" is a powerful statement concerning the power of local action to address global environmental issues. The principle of public participation in environmental decision-making has been formalized in the Aarhus Convention on Access to Information, Public Participation in Decision-making and Access to Justice in Environmental Matters, established by UNECE in 1998 and discussed in Chapter 4, which makes it a legal requirement for signatories to "provide for early public participation, when all the options are open and effective public participation can take place."

Public participation

Participation involves designing institutions and rules that can involve all interested parties in decision-making to produce a consensus that forms the basis for legitimate decisions. The consensual ethos underpinning public participation is rooted in Jürgen Habermas's (1984) philosophy of communicative rationality, which was itself a reaction to the forces of consumerism that he perceived to have alienated citizens from the decisions that shape their lives. Public participation involves consulting with stakeholders on a range of formal environmental management processes, including assessments of environmental risks and impacts, decisions relating to environmental actions and management priorities (Stern and Fineberg 1996). Renn *et al.* (1995: 2) suggest that public participation takes place through all "forums for exchange that are organized for the purpose of facilitating communication between government, citizens, stakeholders, interest groups, and businesses regarding a specific decision or problem." Participation helps frame environmental decisions, for example deciding what types of impact are most important to a community, and can also help identify the most appropriate units for analysis and the assessment methods that are used to capture them.

Of course, involving the full range of people who might be affected in most environmental decision-making processes presents a Herculean task. Stakeholders may be either individuals or organizations, and often perform multiple roles; for example, it is possible to be a resident and an expert at once. Schmitter (2002: 62–63) has distinguished between seven different kinds of actors who may be involved in decision-making:

Right holders: usually covers any citizen or member of the public.

Spatial holders: those who will be affected by their spatial proximity, such as residents.

Share holders: actors who actually own a material element that will be affected by the decision.

Stake holders: those who could affect or be affected by a decision.

Interest holders: any actor who desires to take part in the decision-making process, usually on behalf of some other group.

Status holders: actors who are obliged to take part in decision-making due to some formal responsibility.

Knowledge holders: specialists and experts whose participation is required in order to lend the decision authority.

Because many participation processes are resource intensive they are often relatively small-scale, making it hard to involve the full range of stakeholders in the process. There may also be issues identifying relevant stakeholder groups where new issues are being tackled, or difficulties involving groups who have traditionally been marginalized. Stakeholder analysis provides a tool for identifying who should be involved in decision-making, and how much influence they should have (Grimble and Wellard 1997), and can be used as a network management tool (discussed in Chapter 5). Stakeholder management can involve prioritizing stakeholders by their level of power, or conversely actively seeking to include stakeholders who are usually weaker and more remote. A full list of possible stakeholders can be developed in consultation with key stakeholders, which can then be categorized in order to ensure that a representative range of stakeholders is involved in the final decision-making process (Prell *et al.* 2007). Prioritization is necessary to avoid what De Vivero *et al.* (2008) term the "participation paradox," whereby involving greater numbers of actors actually results in smaller contributions from each, and thus less effective participation.

A key idea in the literature is that public participation must be "fit for purpose," or appropriate to the goals of the process. This can vary from simply providing stakeholders with information concerning a decision that has already been taken, to running residential workshops whereby

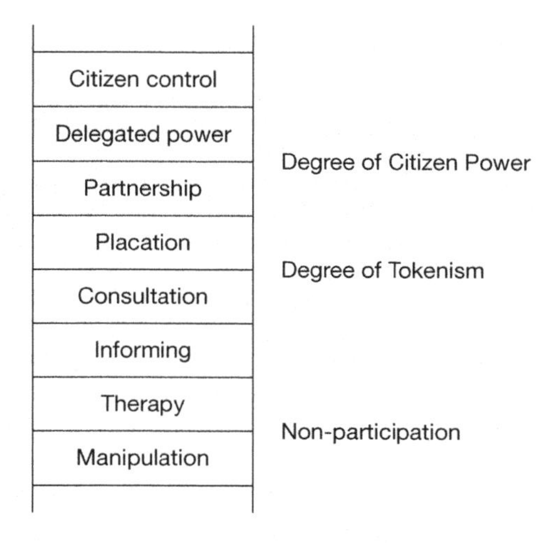

Figure 9.1 ***Ladder of participation***
Source: adapted from Arnstein 1969.

stakeholders deliberate upon a decision and the factors affecting it in great depth. Sherry Arnstein, who pioneered the desegregation of the US health system in the 1960s, developed what is now her classic "ladder of participation" to describe the different levels of participation (Figure 9.1).

At the bottom of the ladder, public participation is driven by an information deficit model, which assumes that the public are largely ignorant about environmental issues (Irwin 1995). Increasing awareness and changing public behavior are seen simply as matter of providing information. But responding to public inaction with "more science" and information does not necessarily work. The public need to trust the sources of science and knowledge; an issue intimately bound up with their beliefs, opinions and experiences of different actors and institutions.

For these reasons, public participation has become more cooperative over time, moving from simply telling the public about decisions to involving them as partners in decision-making (Fischhoff 1995). This transition is captured neatly by the acronyms DAD (Decide–Announce–Defend) and MUM (Meet–Understand–Modify). MUM forms of governance deal with the multiple knowledge claims of different stakeholders by allowing space to discuss them and come to some form of agreement about the decisions at hand (Rydin 2007).

Case study 9.1

Siting an energy-from-waste facility

The county of Hampshire in the UK experienced a waste crisis in the early 1990s, as landfill capacity was limited by the permeable geology of the county, higher regulatory standards were making existing incinerators unviable, and levels of waste were rising (Petts 1995). Original plans to build a new incinerator were defeated by popular opposition in 1992, prompting the county to engage the public in a long-running participation process to develop a more acceptable plan. Exhibitions, questionnaires, road-shows, and media broadcasts were used to recruit participants for three "community advisory fora," each consisting of 16–20 people from diverse backgrounds.

Facilitated by a team of consultants, each group met six times over a half-year period to discuss the available options for the county, detailing their preferred option. The groups received large amounts of information, went on site visits to other facilities in the UK and Europe, received presentations from experts, and eventually agreed on a waste strategy based around three smaller facilities. The private contractor subsequently selected was obliged to engage local communities in each of the proposed areas and used a similar model, this time drawing people from the local communities in which the facilities were to be located. Again, meetings were held, and participants had the opportunity to question experts on a range of potential impacts like traffic, air quality and health, ecology, and visual impact. Engaging directly with the architect, the groups exerted direct influence over the design of the facilities.

The Hampshire case has also been studied in terms of its ability to stimulate social learning, often considered a key outcome of public participation (Petts 2006, Tippett *et al.* 2005). Social learning can be defined as the "process by which changes in the social condition occur—particularly changes in popular awareness and changes in how individuals see their private interests linked with the shared interests of their fellow citizens" (Webler *et al.* 1995). This is obviously a key facet of public participation—community members have personal preferences for different sites based purely upon their own self-interests (no one wants to live near an incinerator, even if it is the "best" place for it). According to Bull *et al.* (2008), it was possible to identify social learning from the experience, as citizens took what they learnt about waste problems and applied it to their own lives, both personally and professionally.

Case study 9.1 describes how community members were involved in siting a waste incinerator, or energy-from-waste facility, in Hampshire, UK.

While there are very good reasons to involve the public in decision-making, it is important to avoid the suggestion that all public knowledge is somehow equally as valid as expert knowledge. Experts have technical knowledge because they devote their professional lives to mastering a specialist field. Wynne (1996) suggests that public knowledge should be used to help frame the ways in which expertise is represented and applied to society, especially in local contexts, but has no role to play in deciding what constitutes expert knowledge or the actual procedures of science. This idea resonates with the risk society thesis, which suggests that the public should have a say in defining what kinds and levels of risk are acceptable. The public may call into question the way in which a decision is being framed, for example, contributing perspectives that might have been missed by experts, or generating more creative approaches to solving problems.

Civic science applies this rationale to the production of scientific knowledge itself, involving the public in ethical decisions concerning what science should and should not be allowed to do (for example, over the use of human embryos in genetic research), and what priorities, fears or concerns exist about new sciences (for example, over nanotechnology). Civic science emerged from high-profile public protests against the deployment of new technologies, like the attempted introduction of genetically modified food to Europe in the 1990s. Such crises highlighted the need to involve the public in the production of scientific knowledge, in order to restore public trust in science, re-orient science towards coping with the complexity of environmental problems and install a democratic element within the governance of science itself (Bäckstrand 2003, Funtowicz and Ravetz 1992).

Problems with public participation

There are a number of challenges to conducting effective public participation:

Asymmetry. Public participation suggests that all stakeholders should be equally engaged, but their stake in the decision may not be equal or comparable. For example, community interests in the siting of a locally unwanted land use (LULU) in their vicinity should outweigh those of

Analytics of governance 9.1

Post-politics

Some theorists argue that modern societies are afflicted by a post-political malaise, whereby people are disenfranchised from the political processes that impact upon their lives. Rather than opening up space for more democratic decision-making as its proponents argue, public participation simply generates consensus between actors who already share the same values and want the same outcomes. Scholars have termed this condition post-political, in that anyone antagonistic to the consensus established by public participation is marginalized as idealistic, impractical or extremist and their speech or actions are not recognized. (There is a certain irony here in that the ideas of Habermas that underpin participatory governance were developed themselves in response to the perceived political alienation produced by consumer society.) Beck (2000: 80) has suggested that today's institutions and bureaucracies act like "zombies—dead long ago but still haunting people's minds"–incapable of capturing real political values and simply going through the motions of consultation and participation.

Within the environmental sphere sustainability is seen as a classic case of post-politics—it is a consensus with which no one can disagree, and yet in practice equates to a continuation of business as usual (Swyngedouw 2007). Sustainability often amounts to little more than development that is geared toward service sector professionals or the aesthetic tastes of middle-class environmentalists, with little consideration given to issues of social justice and political change (Agyeman *et al.* 2003, Krueger and Savage 2007).

One explanation for the post-political condition concerns the subservience of politics to economics under capitalism, a situation encapsulated beautifully in the Chinese communist leader Deng Ziaoping's famous answer when questioned about the relative merits of capitalism and communism. In an ultimately pragmatic statement, he replied that it doesn't matter if the cat is black or white as long as it catches mice. Because a great a part of our lives is structured by capitalist relations of production and consumption it is hard to propose meaningful political alternatives. In relation to climate change, Serbian philosopher Slavoj Zizek (2008) has noted that it is easier to imagine the end of the world than the end of capitalism. The inference is that it is only when we begin to seriously contemplate the second part of this statement, and imagine a different type of world, that we can actually start to address the causes of environmental problems.

the general public. There are different kinds of actors who may be affected by a decision, but reflecting different types of stake in a participation process is challenging.

Expert bias. The culture of institutions and decision-makers is often stuck in the mindset that only experts can answer policy questions. As Harrison *et al.* (1998) note, lay knowledges are often discounted in environmental conflicts. This bias may be accompanied by a culture of secrecy, whereby decisions are taken behind closed doors, and then communicated according to the DAD (Decide–Announce–Defend) model.

Lack of resources. Public participation takes considerable time and money to be effective. In many decision-making contexts it is unclear whose responsibility public participation is. Organizations may also see it as a waste of precious resources or lack the necessary skills and capabilities to do it properly.

But while the practical barriers to public participation are very real, there is an underlying assumption that they are surmountable given sufficient time and resources. Of more fundamental concern perhaps are questions concerning the principle of participation. Critics argue that participatory processes tend to focus on the mundane details of decisions that have largely already been taken, rather than engaging with bigger questions concerning the kinds of future people desire and how they would like to get there. In the worst cases, consultation does not feed into any meaningful decision at all, generating the dreaded "tick in the box" syndrome, whereby organizations conduct public participation simply to fulfill a legal requirement. Far from being an extension of democracy, this has led some authors to regard public participation as its antithesis, a position outlined in Analytics of governance 9.1.

Political activism and alternative futures

The French political philosopher Jacques Rancière (2007) argues that rather than occurring in the spaces created for it by the state, real politics takes place apart from the state, outside of the dominant terms of debate. The environmental sphere is no stranger to this kind of politics, having been characterized by numerous social movements that have risen up to protest at the practices and policies of big business or government.

Eco-activism, a form of direct action undertaken by individuals and groups to achieve political, economic, or social goals, was arguably the starting point for modern environmental NGOs like Friends of the Earth, who protested directly against actions they deemed to be environmentally irresponsible. Modern eco-activism can be traced back to a book written by Edward Abbey, called *The Monkey Wrench Gang*, which follows the exploits of four eco-warriors battling against the forces of modern development that are destroying the environment of their beloved American Southwest. Bulldozers, trains and dams are all in their line of fire, as they wage a guerrilla war from the wilderness while being chased by the police and living off the land. The book inspired the establishment of Earth First!, a real-life direct action environmental group who engage in exactly the same sort of disruptive vandalism and sabotage depicted in the book.

The book offers a fascinating insight into the ways in which the identity of environmentalism (and environmentalists) was far less fixed than it is today. The book's author, Abbey, was nicknamed the "desert anarchist," and his characters eat red meat, own guns, drink beer, discard litter, and drive big cars. Like Abbey, they are fiercely critical of both liberals and conservatives, attacking the behavior of indigenous Indians and the activities of conservation organizations like the Sierra Club.

Eco-activists are often highly inventive in the ways in which they target their enemies, organizing strikes, sit-ins, mechanical sabotage, and property destruction. For example, the Climate Camp movement in the UK establishes temporary settlements to draw both attention to and disrupt existing or proposed sites of greenhouse gas emissions. They have camped at Kingsnorth on the proposed site of a new coal-fired power station, at the existing Drax coal-fired power-station, which is the largest single emitter of greenhouse gases in the UK, and at Heathrow airport.

The tradition of direct protest is closely related to the environmental justice movement, widely considered to have been born in the 1970s at Love Canal, a community in upstate New York whose town was built on an old chemical site. Following a spate of health problems centered on the local school, a local mother, Lois Gibbs, mobilized the community to fight a legal case against Hooker Chemicals and, in a landmark case, won compensation. Lois Gibbs's actions at Love Canal kick-started the environmental justice movement, and she formed the basis for Julia Roberts's character in the film *Erin Brockovitch*. Some

30,000–50,000 sites like Love Canal exist across the USA, and the environmental justice movement has highlighted that environmentally polluting activities are disproportionately located in economically disadvantaged, politically disenfranchised and ethnic minority neighborhoods (Bullard 1990). Similar community resistance has taken place all over the world (Guha and Martinez-Alier 1997), and Case study 9.2 considers one of the best known examples from the developing world.

If there is an antidote to post-politics then it comes in the form of alternative visions that can inspire political action. Fiction plays an important role in this regard. Written by Ernest Callenbach in 1974, the book *Ecotopia* is set in the then mythical future of 1999, in which a new nation made up of Northern California, Oregon and Washington has seceded from the USA in order to construct an ecologically sustainable society. The narrative is woven from the diary entries and reports of William Weston, a newspaper reporter who is sent to investigate the new country, and through his eyes we gradually learn about different aspects of Ecotopian society. While human blood-sports are still relatively foreign to contemporary society, the state-sanctioned use of cannabis is not, and the book offers a fascinating insight into what a green society might feel like to live in, covering issues as diverse as sewage, health, politics, and sex.

Passages such as this, where Weston ventures onto San Francisco's famous Market Street for the first time, demonstrate the power of imagining an alternative vision of the future (Callenbach 1974: 11):

> The first shock hit me at the time I stepped onto the street. There was a strange hush over everything. I expected to encounter something at least a little like the exciting bustle of our cities—cars honking, taxis swooping, clots of people pushing about in the hurry of urban life. What I found when I had gotten over my surprise at the quiet, was that Market Street, once a mighty boulevard striking through the city down to the waterfront, has become a mall planted with thousands of trees. The "street" itself, on which electric taxis, minibuses and delivery carts purr along, has shrunk to a two-lane affair. The remaining space, which is huge, is occupied by bicycle lanes, fountains, sculptures, kiosks and absurd little gardens surrounded by benches.

Ecotopia also explores what a steady-state economy might look like in practice (discussed further in Key debate 9.1), considering such practicalities as energy production, building construction, military strategy, agriculture, defense, education, and medical systems:

> The stable-state concept may seem innocuous enough, until you stop to grasp its implications for every aspect of life, from the most personal to the most general. Shoes cannot have composition soles because they will not decay. New types of glass and pottery have had to be developed, which will decompose into sand when broken into small pieces. Aluminum and other nonferrous metals are largely abandoned, except for a few applications where nothing else will serve—only iron, which rusts away in time, seems a "natural" metal to the Ecotopians. Belt buckles are made of bone or very hard woods. Cooking pots have no stick-free plastic lining, and are usually heavy iron. Almost nothing is painted, since paints must be based either on lead or rubber or on plastics, which do not decompose. And people seem to accumulate few goods like books; they read quite a bit compared to Americans, but they then pass the copies on to friends, or recycle them. Of course there are aspects of life which have escaped the stable-state criterion: vehicles are rubber-tired, tooth fillings are made of silver, some structures are built of concrete, and so on. But it is still an amazing process, and people clearly take great delight in pushing it further and further.

Anthropologist James Holston (1999) argues that imagining different futures is necessary in order to prevent governance simply reinforcing the status quo. American engineer, author, and futurist Richard Buckminster-Fuller puts it another way, saying, "you never change things by fighting against the existing reality. To change something, build a new model that makes the old model obsolete." Alternatives are required in order to steer society in progressive directions. The short-lived Cascadian Independence Movement, which sought to establish an independent Pacific coast state in the Northwest USA after the fashion of *Ecotopia*, is testament to the ability of fiction to inspire reality. In the real world, the World Social Forum stages an event called "Another World Is Possible," first held in 2001 in Porto Alegre in Brazil, which shares visions of an alternative future not based on economic globalization. Porto Alegre is an apt place to hold such an event, as it operates participatory budgeting in each neighborhood, whereby the residents set priorities for local government spending. In an indication of the appetite for alternative visions of society, the event drew 150,000 participants in 2005.

Holston argues that national citizenship has become problematic in the context of massive migrations and loss of shared community, opening up the possibility of multiple citizenships based on urban, local, regional, or transnational affiliations. Increasingly, grassroots movements have sought to create alternative futures at the local level. The Reclaim

Case study 9.2

Chico Mendes and the *seringueiros*

Chico Mendes was born in 1944 on a rubber estate in the Acre region of Northeast Brazil. He began work aged nine, like his father, as a rubber tapper, or *seringueiro*, extracting latex from rubber trees for the estate owners. Life was tough on the estates—schools were not allowed, and the rubber tappers were tied to the estate owners through the debts they owed in return for their equipment. In the 1970s, the military government began an Amazonian occupation process based on agricultural and cattle ranching. This led to the destruction of the Amazon's natural resources and the expulsion of Indians and rubber tappers. In Acre, speculation resulted in the sale of old *seringais* (rubber tapper settlements) to big companies, which began "cleaning" the forest by burning it. Financed by the World Bank, the BR 364 highway, connecting the capital of Acre, Rio Branco, Rondônia and Mato Grosso with the rest of Brazil, facilitated the invasion of the cattle raisers and loggers.

In 1976, Mendes invented a form of resistance called the *empate*—a collective effort to block the action of the loggers in charge of felling trees. In a typical *empate*, a group of 100–200 people would move peacefully into the workers' camp and convince them to lay down their chainsaws. The grassroots resistance of the rubber tappers led to formal engagement with the state system, through the formation of the National Council of Rubber Tappers in 1985. The National Council of Rubber Tappers gained wide support from international environmental organizations for the creation of extractive reserves, which protected areas of forest for traditional extractive workers. As commercially viable ventures, they promised a way to create schools, health centers and cooperatives managed by rubber tappers.

Mendes made many enemies—ranchers, landowners, politicians, local police—and on December 22, 1988, paid the ultimate price, being assassinated by hired gunmen at the age of 44. The legacy of Chico Mendes's actions is considerable: international recognition for the ecological devastation of the Amazon and the plight of its inhabitants in the international environmental community, an internationally recognized institution representing the interests of rubber tappers, and 21 extractive reserves covering some 3.3 million hectares of the Amazonian rainforest.

Key debate 9.1

Steady-state economics

For many, the question of whether markets can be used to address environmental issues is subsidiary to the need for a different kind of economy. As the rapid development of China and India highlights, it does not matter if economic activities become more energy efficient if the overall amount of global emissions doubles (an argument that the USA uses to insist that binding emissions reductions must be applied to developing as well as developed countries). Visions of what a different economy might look like range from advocating low carbon versions of what we currently have, through to questioning the need for growth itself. For example, the recently established field of "happiness economics" suggests that economic wealth does not correlate very well with the levels of life satisfaction experienced by people. Beyond the level where basic needs such as food and shelter are met, people don't report being any happier when they become richer. So within developed countries, the proportion of people who say that they are happy is no higher today than it was 40 years ago, despite the fact that average incomes have increased considerably. Similarly, people in developed countries aren't any happier than people in lower-income countries. This phenomenon is called the "Easterlin Paradox," after a paper published by Richard Easterlin in 1974, which surveyed the reported happiness of people across the USA and a range of developing countries.

Ecological economist Herman Daly (1991) suggests that sustainability requires a shift to a steady-state economy, in which continuous economic growth ceases. Popular fictions of future ecological societies, like Ernest Callenbach's novel *Ecotopia*, are often based upon ideas of a steady-state economy. More recently, steady-state ideas have found a voice in official discussions of sustainability (Jackson 2009, Sustainable Development Commission 2009). Popular movements in France and Italy have taken this a step further, arguing that we actually need de-growth, or *decroissance*, in order to bring Western society back within sustainable limits.

Of course, the problem is that growth is hardwired into our economic system by the monetary system. In capitalist economies, industrial activity is funded by capital that is borrowed at a rate of interest. The system of credit demands constant growth due to the necessity of having to pay interest on borrowed capital. A steady-state economy thus requires an alternative, sustainable, system of money supply, which would be available at no interest. While this is not the place to pursue the monetary details of such an argument, it is worth noting that, even during the financial crisis of 2008, governments were reluctant to bypass the banking system. Despite coming straight from the public purse, money was released into the economy through private banks, who lent it at interest. Another slightly discouraging precedent is that of France, which actually experimented with steady-state policies in the 1970s. The government took seriously the idea that the numbers of hours worked should decrease as technology and efficiency increased. But reducing the hours of public workers met with resistance—it appears that people would rather earn more money than have time off.

the Streets movement of the 1990s, which held illegal street parties on major urban roads across Europe, was simultaneously an expression of alternative counterculture and an event that temporarily brought a different, car-free, space into existence. Arroyo-fest in Pasadena, California, shut eight miles of the 110 freeway in 2002 to allow bikes to ride and pedestrians to walk from York Boulevard in South Pasadena to Sycamore Grove Park in Highland Park. The idea was to celebrate the Arroyo as a historical, cultural and landscape feature of life in Southern California and raise environmental awareness (Gottlieb 2007). Temporary events like this are powerful not only because sections of society literally "reclaim space," but because they plant the seeds of another possible future in the collective consciousness, reinvigorating the broader cultural context of metagovernance that frames first- and second-order governance.

Communities are also engaging in permanent transformations. The Transition Town movement, which encourages "community-led responses to peak oil and climate change, building resilience and happiness" (Hopkins 2008: 8), was founded in southwest England in 2006 by a permaculturalist called Rob Hopkins, and to date there are 277 Transition Initiatives across the world. Transition Towns take charge of their own basic needs in order to make them more resilient to climate change and peak oil, applying the principle of localization to food production, energy production, building materials and waste. The movement is very much about action, and transition involves making physical changes to the places in which people live. Communities develop their own interventions; as the founder states, "it makes top-down solutions almost redundant. . . . resilience-building is about working on small changes to lots of niches in the place, making lots of small interventions rather than a few large ones" (Hopkins 2008: 55). A role is acknowledged for the state—communities can't go it alone—but they shouldn't wait for the state to take the lead.

While the Transition Town movement has been criticized for becoming overly hierarchical (people have to pay to attend courses in order to have their town officially recognized by the movement) (Smith 2010), grassroots environmental movements increasingly understand that effecting material changes in the real world can prompt political change. Gandhi used tactics like encouraging Indians to spin their own cloth and make their own salt to help drive the British out of colonial India. Simple acts like reclaiming unused public spaces for community gardens can help generate a sense of place and community. City Repair in Portland, Oregon, utilizes this idea by reclaiming urban spaces to create community-oriented places. Taking charge of public space, they argue, engenders greater neighborhood communication and empowerment. As they say, "streets are usually the only public space we have in our neighborhoods. But most all [*sic*] of them have been designed with a single purpose in mind: moving cars around" (City Repair 2010). Their Intersection Repair initiative encourages the reclamation of intersections as public squares for the whole community, changing what it looks like and how it is used (Plate 9.1).

As their website states (ibid.),

> one neighborhood may paint a giant mural on the intersection and stop there. Another may go through many phases: painting the street, installing a community bulletin board, building a mini-cafe on a corner, reconstructing the intersection with brick and cobblestones, opening businesses to make it a village center . . . and on and on!

Plate 9.1 ***Painted intersection in Portland, USA***
Source: reproduced with permission from City Repair.

Like Chico Mendes and the rubber tappers discussed in Case Study 9.2, City Repair has begun to have an impact upon formal channels of governance, as the City of Portland has now passed a planning ordinance that allows for Intersection Repair paintings.

By way of concluding this section, it is worth noting that the capacity to engage is itself far from universal. As Julian Agyeman (2005: 105–6) states, "grassroots environmental justice groups are often lacking in their ability to frame the issue, seize on political opportunities, and mobilize the political and financial resources need to be more proactive, that is, heading off problems before they arise." It is no coincidence that Edward Abbey, Ernest Callenbach and Friends of the Earth all emerged at around the same time in the same place—1960s California—and that City Repair and Portland today lie in the heart of what would have been Ecotopia. Ernest Callenbach was directly influenced by Edward Abbey, and both were influenced by the wider countercultural movements associated with the hippy movement of Ken Kesey. This unique confluence of forces powerfully demonstrates the different traditions and capacities for environmental action that places enjoy.

Conclusions

Involving communities in governing the environments in which they live makes intuitive sense—it enhances the legitimacy of decision-making, is fairer to those affected and potentially more effective than trying to impose external controls. Further, while governance steers society, it does not dictate the direction in which it should be steered, necessitating some form of political engagement. Public participation seeks to engage the public formally with environmental governance at all levels, and has become part and parcel of decision-making processes in the developed world.

Table 9.1 lists the key strengths and weaknesses associated with participation. For the most part, these surround the democratic status of participation, and specifically whether it really allows people to have a meaningful say in decisions that are taken. Certainly, the promise of participation is that it opens up previously closed decision-making processses to public participation, enhancing the quality and legitimacy of the resulting decision. Conversely, there are times that public participation has little meaningful impact upon a final decision, or simply reinforces the status quo. The consensual model of participation

Table 9.1 ***Strengths and weaknesses of participation***

Strengths	*Weaknesses*
Opens up decision-making to democratic involvement	Difficulties of involving all stakeholders
Can be used to support any decision-making process	Costly and time consuming to do
Improves the quality of decisions that are made	Little meaningful impact upon key decisions
Enhances legitimacy of decision-making	Consensus prevents dissenting voices from being heard

in particular has been criticized for excluding opinions lying outside of the mainstream.

In opposition to formal channels of participation, grassroots movements and eco-activism seek to make more radical viewpoints heard. Indeed, the environmental movement was (and continues to be) characterized by protest movements of this kind. The post-political critique splinters democracy into formal engagement, which takes place inside the system, and informal protest, which takes place outside. But while this is an appealing dichotomy, the lines are rarely so clear in practice. Public participation blurs the boundaries between expert and non-expert, public and private, and citizen and government. Informal grassroots groups can become institutionalized over time and engage in formal governing processes, like Friends of the Earth or the *seringueiros*. At the level of metagovernance, the world beyond the formal channels of environmental governance enriches it with new values, innovative ideas, and dynamic institutions.

Questions

- Using examples, discuss whether public participation is a deficient form of democracy.
- Think of an alternative vision of society (it can be drawn from the popular media, broader cultural sources, or actual social movements). What implications does it hold for environmental governance?

Key readings

- Beck, U. (1992b) "From industrial society to the risk society: questions of survival, social-structure and ecological enlightenment," *Theory, Culture, Society*, 9: 97–123.
- Renn, O. (1999) "A model for an analytic deliberative process in risk management," *Environmental Science and Technology*, 33: 3049–55.
- Swyngedouw, E. (2007) "Impossible sustainability and the post-political condition," in R. Krueger and D. Gibbs (eds) *The Sustainable Development Paradox: Urban Political Economy in the US and Europe*, New York: Guilford Press, 13–40.

Links

- http://tcktcktck.org/. This site is a major hub for mobilizing global civil society support for an agreement on climate change.
- http://climategroundzero.net/. UK-based climate activism group.
- http://top-lists.info/green-roofs-future-of-the-world. Blog with pictures of what the world might look like with green roofs—an example of the power of envisioning an alternative, greener city.

10 Conclusions

Intended learning outcomes

At the end of this chapter you will be able to:

- **Summarize the evolution of governance in the environmental field.**
- **Assess the key strengths and weaknesses of different modes of environmental governance.**
- **Understand emerging themes in environmental governance.**

Introduction

In its most basic form, governance involves actors beyond the state in the practice of governing by securing the conditions to enable collective action. This concluding chapter reconsiders the evolution of environmental governance, providing a brief summary of each chapter. It then compares how each mode of governance facilitates collective action, assessing their strengths and weaknesses. It finishes by presenting eight hypotheses on environmental governance that are intended to prompt discussion, and identifies emerging themes for those seeking to pursue study in this area.

Environmental governance reconsidered

While governance is a relatively recent phenomenon, the practice of governing more broadly has a long history, accompanying the emergence of the modern nation state, which required an administrative government to tend to its population. The transition from government to governance that has occurred over the last 25 years has been a gradual process, in which various roles that were traditionally performed by government have been opened up to actors beyond the state.

Depending on the political viewpoint, opening up the process of governing in this way can be seen as an extension of democracy that strengthens decision-making, or conversely as part of a wider undermining of the state and public sector by neoliberalism and economic globalization. As with most things in life, these positions are both true, insofar as they tell part of the story. Governance is best seen as a political response to the set of conditions that emerged in the late 1980s, in which economic considerations became the dominant drivers of politics. It was not as if the goals of government suddenly changed, but rather that they required new ways of achieving them within the context of economic globalization.

While the shift from government to governance has not been restricted to the environmental domain, the complex nature of environmental problems suited them to the governance approach. High levels of uncertainty surrounding environmental change, the global or transboundary nature of many problems, and the lack of global institutions to make and enforce decisions, creates an obvious need to include broader sets of actors in the process of decision-making. Perhaps more than most areas of government, the environment emerged as an object of governance primarily at the global level, reflected by the profusion of institutions like NGOs and international bodies.

The story of global environmental governance can be told as one of excitement and agenda-setting in the early days, followed by an increasing recognition of the difficulties of securing legally binding international agreements and implementing them. The sheer number of states involved in negotiations provides a structural limitation to the current system of international agreements. Added to this, economic forces are now critical in determining the viability of actions to address environmental issues. For example, imposing levies on unsustainable imports is effectively a legal question for the WTO, not UNEP. Such a fragmented system of political jurisdictions makes it hard to address global environmental systems, but an overarching global government or environmental enforcement body is neither desirable nor likely. Given these structural limitations to multilateral action, networks and markets are increasingly heralded as the most promising means through which to implement environmental agreements.

Network governance involves groups of actors with common interests coming together to work towards mutually beneficial outcomes, leveraging the collective resources of diverse actors and simply bypassing reluctant national governments. The profusion of NGO and QUANGO networks that have sprung up to help put the agreements reached at Rio, Kyoto and Johannesburg into practice is the most exciting and dynamic development in the field of environmental governance, generating real changes to the political and economic behavior of governments, companies and the public. The fact that networks achieve change voluntarily attests to their power to coordinate action, although questions surround the effectiveness, accountability, and legitimacy of network governance.

Many of these networks have emerged to support market mechanisms. Market approaches promise to address environmental problems efficiently by allocating resources through the laws of supply and demand. That said, the armies of bureaucrats, scientists, and eco-entrepreneurs required to create markets in environmental goods like carbon make it hard to determine whether they really are either efficient or capable of changing the status quo. What is clear is that markets do not operate in a vacuum but within parameters set by the state—the question is rarely "market or no market," but rather what role markets should play as part of a mix of governance approaches.

Transition management seeks to steer the economy by encouraging low carbon innovations. In this case, the ability of the state to manipulate economic conditions is critical, although the approach has been criticized for underplaying the importance of wider social expectations and the policy context. Transition management gets less bogged down in the ethical debates that hamstring many approaches to the environment, seeking to transform society by changing its material basis.

Adaptive governance makes society more responsive to changing environmental conditions. The ability to adapt is dependent on designing institutions and decision-making procedures that are capable of experimenting and learning from the social and ecological impacts of different interventions. Emergent modes of environmental governance like transition management and adaptive governance are more experimental in their approaches. While the modes of governance considered in the second half of the book clearly relate to one another, by way of conclusion it is useful to compare them to one another, assessing their key characteristics, and teasing out their strengths and weaknesses.

Comparing the different modes of environmental governance

Table 10.1 lists seven categories for each mode of governance. The first five relate to their characteristics, while the subsequent two relate to their cost and ability to steer. These categories are not intended to be either complete or definitive, but build upon the key features of governance identified in the first four chapters (the importance of collective action, rules and institutions) and the individual conclusions at the end of each chapter.

Geographically speaking, networks are highly flexible because they do not require a common regulatory or political framework in which to operate. They work topologically, connecting nodes rather than enclosing space, which allows them to form global (or transnational) connections. By comparison, market and transition modes require coherent regulatory frameworks within which to operate, which need to be framed either by the state, or agreed between states. When expanded, both modes are vulnerable to "leakage," whereby undesirable activities simply relocate outside of the regulated area. Adaptive governance tends to work best when applied to specific processes, which might involve a single social–ecological system. This requirement makes these modes more suited to regional and local governance problems.

Table 10.1 ***Comparing the different modes of governance***

	Network	*Market*	*Transition*	*Adaptive*
Scale	Transnational	National/ international	National/ city	Local/ regional
Source of rules	Network	State-led partnerships	State-led partnerships	Network
Requirements for collective action	Capacity	Regulations	Regulations	Capacity
Status of actors	Stakeholders	Producers and consumers	Innovators and adopters	Stakeholders
Role of institutions	Facilitating (common goals)	Regulating	Managing	Learning
Cost	Low	Low	High	Medium
Potential to steer	Medium	Medium	High	Medium

As Table 10.1 shows, the source of rules for each mode reflects these geographical constraints. Rules governing markets and transition management are set largely by the state, or the state working in partnership with industry, while the actors involved in network and adaptive modes set their own rules. The requirements for collective action follow similar groupings, with markets and transition approaches requiring regulatory frameworks in which to operate, compared to network and adaptive modes that work on the basis of their own capacity to act, in terms of the resources and knowledge present in a network. Accordingly, network and adaptive modes cast actors as stakeholders, who are involved because they have some form of interest. Actors in the market mode are cast as producers and consumers, while in the transition mode they are cast as innovators and adopters. These roles are again similar, with the market mode emphasizing the financial aspects of economic exchange, and transition management emphasizing the knowledge-driven element of economic growth.

In terms of the institutional qualities demanded, policy-makers are most comfortable with the market and transition modes, as regulation and management are more familiar activities. Indeed, transition management can be seen as a strategic form of market governance. Networks require softer skills of facilitation and encouragement, generally working better when they do not involve a strong state presence. The adaptive mode can be seen as an extension of network governance to include ecological elements, although it requires a more experimental approach that requires decision-makers to accept failure as part of resource management. That said, more established modes of governance also entail degrees of risk and uncertainty. Market advocates are at pains to point out the process of designing markets proceeds by trial and error—they rarely work perfectly straight away. Similarly, because networks can proliferate quickly around an issue, many will inevitably fail or become irrelevant and simply cease to exist over time. The ability to experiment, learn, and potentially fail, is an important characteristic of all governance modes. Social learning that accrues through participation represents a form of collective learning, while the co-evolution of technology and society that characterizes transition management is a form of collective experimentation.

In terms of cost, it is clear to see why networks and markets have proven so popular, promising low-cost ways in which to address environmental issues (whether they actually are low cost is debatable, but this is certainly the claim). Transition management, while costlier, appeals to policy-makers because it promises a strong steer towards the kind of systemic transformation that is required to achieve a low carbon economy. Similarly, adaptive governance is attractive to policy-makers charged with maintaining resources and services in the face of changing environmental conditions and a highly unstable political funding environment.

Eight hypotheses

Much of the discussion in this book has tended to be rather circumspect, seeking to present a balanced view of different aspects of environmental governance. By way of conclusion, eight more strident and provocative hypotheses are presented that are intended to prompt discussion.

Networks and markets are the best things that we have. Environmental problems are not going to disappear, and neither is state sovereignty or capitalism. A binding global agreement on carbon emissions, which includes all nations, is probably not going to happen—there is no political appetite for binding regulation and the system of international relations is structurally flawed. This doesn't necessarily matter though —all major emitters are preparing emissions trading schemes, and voluntary networks for carbon reduction are proliferating. Further, as the IPCC is now increasingly accepting, the world will not meet its emissions reductions targets, meaning that soft geoengineering options, like tree planting, will be required in order to remove CO_2 from the atmosphere. In the context of overshoot, mechanisms that rely on network governance for implementation, like the CDM and REDD+, will therefore become increasingly critical. Although networks have been criticized for their voluntary basis, and markets for exacerbating existing inequalities, they are the best things that we currently have, so we should work with them.

Governance is about evolution, not revolution. Many modes of governance are based upon the idea that there are different levels at which governance unfolds. For example, the three orders of first, second and meta-governance identified in Chapter 2 are mirrored in the niche, regime and landscape levels of transition management, and the nested hierarchy of adaptive governance. Orders not only suggest that there are different spatial scales at which governance unfolds, but also different temporal scales. Change at the level of meta-governance requires long-term shifts in cultural attitudes and political opinions. Lower orders of governance affect and are affected by this level, but only slowly.

Getting the mix of approaches right is critical. The current governance landscape reflects a diversity of approaches and institutions that have emerged to deal with environmental issues. On the downside, this makes it difficult to attribute accountability and measure the effectiveness of governance (Bulkeley and Newell 2010). But on the upside, imposing a single mode of governance would probably be counter-productive as the problems are simply too complex. There is no magic bullet for solving environmental issues because the problems and potential solutions vary greatly. The different strengths and weaknesses of governance modes identified above makes each suited to different places, scales and problem sets. A key priority for policy-makers involves creating the right mix of governance approaches, which may also involve traditional forms of regulation.

For example, in his book, *Kyoto 2*, Oliver Tickell (2008) advocates a carbon tax on emissions at the point of extraction—in other words, when coal or oil is actually taken out of the ground. The costs would then be passed upwards through the commodity chain and the proceeds could fund a "Sky Trust," which would be used to fund mitigation and adaptation in the developing world. While politicians fear that this kind of alternative is too blunt and interventionist, risking excessive disruption to Western economies (which depend on cheap oil), there is no reason in principle why such a scheme could not work alongside emissions trading schemes, although the practicalities of achieving political and corporate buy-in return us to the question of governance.

Governance requires political vision. In order to steer, a society needs to have goals. While governance steers society, it does not tell us in which direction we should steer. Participation enhances the legitimacy of decision-making, making it fairer to those affected and potentially more effective than simply imposing external decisions. While participation is costly, and requires decision-makers to loosen their grip on power in order to allow the public to meaningfully affect a decision, it has a critical role to play in generating a shared vision concerning the direction society should be steered in.

In focusing on rules and procedures, governance has been criticized for being post-political and neglecting bigger questions concerning the kind of future that is desirable. Despite the one-world discourse, progress in shifting allegiances from individual nations to the planet has been painfully slow. Further, the tendency of the one-world ideal to exclude certain groups prompts the question of whether it is a useful vision.

Although it often isn't viewed as such, governance can be seen as a source of new identities and political visions, around, for example, cities or socio-ecological systems. In this way, governance prompts us to think about social and political innovation as well as technological innovation.

Governance is about learning. The success of governance depends on the ability to adapt to changing contexts through a process of learning. Both science and capitalism have problems dealing with non-linear changes, as cost-benefit analysis and traditional forms of resource management are based on the concepts of equilibrium and engineering resilience. With no stable nature to tell us what to do, these standard managerial approaches to intervention are inappropriate. In order to govern against a shifting backdrop of economic, political and environmental change without losing sight of political and social goals, governance requires institutions that can learn. This involves setting rules that allow for experimentation and transformation, and which recognize failure as an inherent part of the learning process.

The ability to learn and change is common across the different modes of governance, and is potentially inherent in the concept of governance itself. As the product of late modernity (i.e. the second half of the twentieth century), governance echoes Beck and Beck-Gernsheim's (2001) observation that we now live reflexively, not habitually. In other words, we live through a process of constant reflection on our actions rather than through a simple repetition of actions to produce set end results. The resource costs associated with learning offer a potentially far greater role for universities to play in activities like monitoring, and NGOs in knowledge exchange.

Duality of structure is critical. The tendency of governance to set common goals that allow different actors to devise the most suitable ways to reach them is based upon a duality of structure, whereby small-scale freedom is framed within a large-scale structure. In order to achieve widespread change, networks need to be empowered to act in order to address common goals, or, as Nabeel Hamdi (2004) puts it, scaling up requires scaling down. A critical question for environmental governance involves deciding what form this duality should take. For example, how much small-scale freedom it is desirable or possible to facilitate while still allowing actions to be sufficiently coordinated, and what form should large-scale structures assume. Should an overarching body have enforcement and monitoring duties, should it set and promulgate a common vision, or should it simply provide a platform to share knowledge? The answer to these questions depends on the mode or mix of modes that are chosen, and will to a large extent determine the design of institutions to govern them.

Governments matter. Governments shape markets, innovation contexts, political visions, and legitimacy through the policies that they enact. Given the scale and speed of change required to address climate change, commentators are increasingly advocating direct government action, for example to pump money directly into research rather than incentivizing the market to deliver the right innovations through taxes and subsidies (Lomborg 2007). A simple levy on carbon intensive activities could be used to fund the development of clean technologies, rather than to force behavior change (Galiana and Green 2009). Further, it is possible that network governance might provide a stepping stone towards regulatory change. For example, once the voluntary carbon accounting procedures of a network like the Carbon Disclosure Project become sufficiently widespread they will effectively have secured the support to change the law and make such reporting a legal requirement.

Re-theorizing the state in the context of these strategic questions represents a major challenge for environmental governance, given that many environmental theories lacked a proper theory of how the state worked in the first place. States have the ultimate capacity to shape the structure, mix and institutional context within which governance takes place. Set within the broader context of the 2008 financial crisis and the rise of Chinese capitalism, the idea that the state has an important role to play in governing is back in vogue. That said, the state is as under-resourced as it has ever been, so will continue to work through partnerships and networks to achieve its strategic goals. Understanding the strategic and daily involvement of the state in environmental governance represents a critical challenge.

Hybrid institutions are critical in coordinating action across sectors. Hybrid institutions play a critical role linking environmental action across different sectors of concern. Eco-financial institutions like green banks and infrastructure bonds are facilitating climate change mitigation and adaptation by linking investors to environmental projects. Compliance and monitoring for initiatives like the CDM, REDD, and CSR are driving the emergence of a plethora of institutions that represent civil society, mobilize expert knowledges and wield scientific legitimacy. Hybrid institutions like the IPCC have been vital in linking scientific and political networks, allowing climate change to escape from the ghetto of scientific and environmentalist concern. The conditions under which hybrid institutions form and operate represent an exciting focus for environmental governance, as it is only through such institutions that action to address environmental issues can be coordinated across different sectors.

The future

This book has sought to highlight the potential areas of environmental governance that make it of importance and interest. While the analytics of governance discussed throughout the book, and the emerging priorities identified above, are not exhaustive, they indicate an exciting agenda for environmental governance going forward. As noted in the first chapter, governance is simultaneously heralded as the only way to govern in an unruly fragmented world, and denigrated by others as a corrupted form of politics that simply maintains the status quo. Hopefully, it has become apparent that these positions are not only both partially true, but represent two sides of the same coin. The future under environmental governance is far from certain, but in some ways, that is the point: governance is about steering and emergence, not rigid control and revolution.

Governance has the potential to link people, places and things together in radical new ways. Innovations, whether they take the form of new technologies, novel social networks, or creative political systems, can cause a ripple effect, whereby small interventions can have very large impacts. Breaking with the existing status quo requires diversity, open-mindedness and the capacity to learn and change. In doing these things, governance can help forge new identities and visions for the world in which we want to live.

Bibliography

Abbey, E. (1975) *The Monkey Wrench Gang*, New York: HarperCollins.

Adger, N. (2000) "Social and ecological resilience: are they related?," *Progress in Human Geography*, 24: 347–64.

—— (2010) "Addressing barriers and social challenges of climate change adaptation," in National Research Council, *Facilitating Climate Change Responses: A Report of Two Workshops on Knowledge from the Social and Behavioral Sciences*. Committee on the Human Dimensions of Global Change, Division of Behavioral and Social Sciences and Education. Washington, DC: The National Academies Press, 79–84.

Agrawal, A. (1995) "A southern perspective on curbing global climate change," in S. Schneider and A. Escobar (eds) *Encountering Development*, Princeton, NJ: Princeton University Press.

—— (2005) *Environmentality: Technologies of Government and Political Subjects*, Delhi: Duke University Press.

Agrawal, A. and Narain, S. (1990) *Global Warming in an Unequal World: A Case of Environmental Colonialism*, New Delhi: Centre for Science and Environment.

Agrawal, A., Narain, S. and Sharma, A. (eds) (1999) *Global Environmental Negotiations 1: Green Politics*, New Delhi: Centre for Science and Environment.

Agrawala, S. (1999) "Early science–policy interactions in climate change: lessons from the Advisory Group on Greenhouse Gases," *Global Environmental Change*, 9: 157–69.

Agyeman, J. (2005) *Sustainable Communities and the Challenge of Environmental Justice*, New York: New York University Press.

Agyeman, J., Bullard, R. and Evans, B. (2003) *Just Sustainabilities: Development in an Unequal World*, Cambridge, MA: MIT Press.

Almond, G. (1988) "The return to the state," *American Political Science Review*, 82: 853–74.

Anderson, K. and Richards, K. (2001) "Implementing an international carbon sequestration program: can the leaky sink be fixed?," *Climate Policy*, 1: 173–88.

Anderson, T. and Leal, D. (1991) *Free Market Environmentalism*, Boulder, CO: Westview Press. 2nd revised edn 2001, London: Palgrave Macmillan.

Andersson, R. (2007) *The Politics of Resilience: A Qualitative Analysis of Resilience as an Environmental Discourse, Essay from Stockholm University*. Online. Available HTTP: www.essays.se/essay/9f8729269b/ (accessed 16 December 2009).

Andonova, L., Betsill, M. and Bulkeley, H. (2009) "Transnational climate governance," *Global Environmental Politics*, 9: 52–73.

Andonova, L. and Levy, M. (2003) "Franchising global governance: making sense of the Johannesburg type II partnerships," in S. Stokke and O. Thommessen (eds) *Yearbook of International Cooperation on Environment and Development*, London: Earthscan, 19–32.

Ansell, C. and Gash, C. (2008) "Collaborative governance in theory and practice," *Journal of Public Administration Research and Theory*, 18: 543–71.

Armitage, D. (2010) *Adaptive Capacity and Environmental Governance*, Berlin: Springer Verlag.

Arnstein, S. (1969) "The ladder of Citizen Participation," *Journal of the Institute of American Planners*, 35: 16–24.

Arrow, K. (2007) "Global climate change: a challenge to policy," *Economists' Voice*, 4(3) (June): 1–5.

Attaran, A. (2005) "An immeasurable crisis? A criticism of the Millennium Development Goals and why they cannot be measured," *PLoS Medicine*. Online. Available HTTP: www.ncbi.nlm.nih.gov/pmc/articles/PMC1201695/ (accessed 8 August 2010).

Auld, G., Bernstein, S., Cashore, B. and Levin, K. (2007) "Playing it forward: path dependency, progressive incrementalism, and the 'super wicked' problem of global climate change." Paper presented at the International Studies Association Convention, Chicago, 28 February–3 March.

Axelrod, R., Vig, N. and Schreurs, M. (2005) "The European Union as an environmental governance system," in N. Vig and R. Axelrod (eds) *The Global Environment: Institutions, Law and Policy*, London: Earthscan, 72–97.

Bachram, H. (2004) "Climate fraud and carbon colonialism: the new trade in greenhouse gases," *Capitalism, Nature, Socialism*, 15: 1–16.

Bäckstrand, K. (2003) "Civic science for sustainability: reframing the role of experts, policy maker and citizens in environmental governance," *Global Environmental Politics*, 3: 24–41.

—— (2004) "Scientisation vs. civic expertise in environmental governance: ecofeminist, ecomodern and postmodern responses," *Environmental Politics*, 13: 695–714.

—— (2008) "Accountability of networked climate governance: the rise of transnational climate partnerships," *Global Environmental Politics*, 8: 74–102.

Bäckstrand, K. and Lövbrand, E. (2006) "Planting trees to mitigate climate change: contested discourses of ecological modernization, green governmentality and civic environmentalism," *Global Environmental Politics*, 6: 50–75.

Bailey, S. (1993) "Public choice theory and the reform of local government in Britain: from government to governance," *Public Administration*, 8: 7–24.

Bakker, K. (2005) "Neoliberalizing nature? Market environmentalism in water supply in England and Wales," *Annals of the Association of American Geographers*, 95: 542–65.

Banerjee, S. (2008) "CSR: the good, the bad and the ugly," *Critical Sociology*, 34: 51–79.

Barbier, E. (2010) *A Global Green New Deal: Rethinking the Economic Recovery*, Cambridge: Cambridge University Press.

Bear, C. and Eden, S. (2008) "Making space for fish: the regional, network and fluid spaces of fisheries certification," *Social and Cultural Geography*, 9: 487–504.

Beck, U. (1992a) *Risk Society: Towards a New Modernity*, London: Sage.

—— (1992b) "From industrial society to the risk society: questions of survival, social-structure and ecological enlightenment," *Theory, Culture, Society*, 9: 97–123.

—— (2000) *The Cosmopolitical Perspective: Sociology in the Second Age of Modernity*, Oxford: Blackwell.

—— (2007) "The political and social construction of risk, according to Ulrich Beck," lecture given to Intercultural Dynamics Programme of the CIDOB Foundation, Barcelona. Online. Available HTTP: www.cidob.org/en/noticias/dinamicas_interculturales/la_construccion_politica_y_social_del_riesgo_segun_ulrich_beck (accessed 25 December 2010).

Beck, U. and Beck-Gernsheim, E. (2001) *Individualization: Institutionalized Individualism and its Social and Political Consequences*, London: Sage.

Bennett, M., James, P. and Klinkers, L. (1999) *Sustainable Measures: Evaluation and Reporting of Environmental and Social Performance*, Sheffield: Greenleaf.

Bennett, W. (2002) *News: The Politics of Illusion*, New York: Longman.

Benson, M. (2010) "Regional initiatives: scaling the climate response and responding to conceptions of scale," *Annals of the Association of American Geographers*, 100: 1025–35.

Berkes, F. (1986) "Marine inshore fishery management in Turkey," in National Research Council, *Proceedings of the Conference on Common Property Resource Management*, Washington, DC: National Academy Press, 63–83.

—— (2004) "Rethinking community based conservation," *Conservation Biology*, 18: 621–30.

Berkes, F., Colding, J. and Folke, C. (eds) (2003) *Navigating Social-Ecological Systems*, Cambridge: Cambridge University Press.

Berkes, F., Folke, C. and Colding, J. (eds) (2001) *Linking Social-Ecological Systems*, Cambridge: Cambridge University Press.

Bernstein, S. and Cashore, B. (2004) "Non-state global governance: is forest certification a legitimate alternative to a global forest convention?," in J. Kirton and M. Trebilcock (eds) *Hard Choices, Soft Law: Voluntary Standards in Global Trade, Environment and Social Governance*, Aldershot: Ashgate, 33–63.

Berry, M. and Rondinelli, D. (1998) "Proactive environmental management: a new industrial revolution," *The Academy of Management Executive*, 12: 38–50.

Betsill, M. and Bulkeley, H. (2004) "Transnational networks and global environmental governance: the cities for climate protection program," *International Studies Quarterly*, 48: 471–93.

Betsill, M. and Corell, E. (eds) (2008) *NGO Diplomacy: The Influence of Nongovernmental Organizations in International Environmental Negotiations*, Cambridge, MA: MIT Press.

Beveridge, R. and Guy, S. (2005) "The rise of the eco-preneur and the messy world of environmental innovation," *Local Environment*, 10: 665–76.

Bevir, M. and Rhodes, R. (1999) "Studying British government: reconstructing the research agenda," *British Journal of Politics and International Relations*, 1: 215–39.

Biermann, F. (2001) "The emerging debate on the need for a World Environmental Organization: a commentary," *Global Environmental Politics*, 1: 45–55.

—— (2005) "The rationale for a world environmental organization," in F. Biermann and S. Bauer (eds) *A World Environment Organization: Solution or Threat for Effective International Environmental Governance?*, Hampshire, UK and Burlington, VT: Ashgate Publishing.

—— (2007) "'Earth system governance' as a crosscutting theme of global change research," *Global Environmental Change*, 17: 326–37.

Biermann, F. and Pattberg, P. (2008) "Global environmental governance: taking stock, moving forward," *Annual Review of Environment and Resources*, 33: 277–94.

Biermann, F., Pattberg, P. and Zelli, F. (2010) *Global Climate Governance Beyond 2012*, Cambridge: Cambridge University Press.

Blowfield, M. and Murray, A. (2008) *Corporate Responsibility: A Critical Introduction*, Oxford: Oxford University Press.

Boden, T., Marland, G. and Andres, R. (2010) *Global, Regional, and National Fossil-Fuel CO2 Emissions*, Oak Ridge, TN: Carbon Dioxide Information Analysis Center, Oak Ridge National Laboratory, US Department of Energy.

Bohringer, C. (2003) "The Kyoto Protocol: a review and perspectives," *Oxford Review of Economic Policy*, 19: 451–66.

Börzel, T. and Thomas, R. (2005) "Public-private partnerships: effective and legitimate tools of international governance?," in E. Grande and L. Pauly (eds) *Reconstituting Political Authority: Complex Sovereignty and the Foundations of Global Governance*, Toronto: University of Toronto Press.

Boyd, E., Hultman, N., Roberts, T., Corbera, E., Ebeling, J., Liverman, D., Brown, K., Tippmann, R., Cole, J., Mann, P., Kaiser, M., Robbins, M., Bumpus, A., Shaw, A., Ferreira, E., Bozmoski, A., Villiers, C. and Avis, J. (2007) "The clean development mechanism: an assessment of current practice and future approaches for policy." Working Paper 114, Manchester: Tyndall Centre for Climate Change Research.

Boykoff, M. (2007) "From convergence to contention: United States mass media representations of anthropogenic climate change science," *Transactions of the Institute of British Geographers*, 32: 477–89.

Bradach, J. and Eccles, R. (1991) "Price, authority and trust: from ideal types to plural forms," in G. Thompson, J. Frances, R. Levacic and J. Mitchel (eds) *Markets, Hierarchies and Networks: The Coordination of Social Life Markets*, London: Sage Publications, 277–92.

Brand, R. (2005) "The citizen innovator," *The Innovation Journal*, 10: 9–19.

Bridge, G. and Jonas, A. (2002) "Governing nature: the reregulation of resource access, production, and consumption," *Environment and Planning A*, 34: 759–66.

Brockington, D. (2009) *Celebrity and the Environment: Fame, Wealth and Power in Conservation*, London: Zed.

Brown, J. and Purcell, M. (2005) "There's nothing inherent about scale: political ecology, the local trap, and the politics of development in the Brazilian Amazon," *Geoforum*, 36: 607–24.

Browne, P. (2010) "China's Copenhagen paradox," *Inside Story*, January. Online. Available HTTP: http://inside.org.au/chinascopenhagenparadox/ (accessed 13 September 2010).

Buckley, N., Mestelman, S. and Muller, A. (2005) "Baseline-and-credit style emission trading mechanisms: an experimental investigation of economic inefficiency," *Climate Policy*, 3: 42–61.

Bugler, W., Hangartner, C., Jenkinson, C. and Underwood, J. (2010) "Transnational climate change networks," Environmental Governance Series Report 2, School of Environment and Development, University of Manchester, Manchester, UK.

Bulkeley, H. (2005) "Reconfiguring environmental governance: towards a politics of scales and networks," *Political Geography*, 24: 875–902.

Bulkeley, H. and Betsill, M. (2003) *Cities and Climate Change: Urban Sustainability and Global Environmental Governance*, London: Taylor & Francis.

—— (2004) "Transnational networks and global environmental governance: the Cities for Climate Protection program," *International Studies Quarterly*, 48: 471–93.

Bulkeley, H. and Kern, L. (2006) "Local government and climate change governance in the UK and Germany," *Urban Studies*, 43: 2237–59.

Bulkeley, H. and Moser, S. (2007) "Responding to climate change: governance and social action beyond Kyoto," *Global Environmental Politics*, 7: 1–10.

Bulkeley, H. and Newell, P. (2010) *Governing Climate Change*, London: Routledge.

Bull, R., Petts, J. and Evans, J. (2008) "Social learning from public engagement: dreaming the impossible?," *Journal of Environmental Planning and Management*, 51: 701–16.

Bullard, R. (1990) *Dumping in Dixie: Race, Class and Environmental Quality*, Boulder, CO: Westview Press.

Bumpus, A. and Liverman, D. (2008) "Accumulation by decarbonisation and the governance of carbon offsets," *Economic Geography*, 84: 127–55.

Burkeman, O. (2011) "SXSW 2011: the internet is over," *Guardian*, 15 March.

Cadbury, A. (2002) *Corporate Governance and Chairmanship: A Personal View*, Oxford: Oxford University Press.

Caldeira, T. and Holston, J. (2005) "State and urban space in Brazil: from modernist planning to democratic interventions," in A. Ong and S. Collier (eds) *Global Anthropology: Technology, Governmentality, Ethics*, London: Blackwell, 393–416.

Callenbach, E. (1974) *Ecotopia*, Berkeley, CA: Banyan Tree Books.

Carbon Trust (2006) *Frameworks for Renewables*. Online. Available HTTP: www.carbontrust.co.uk/publications/publicationde (accessed 8 November 2010).

Carpenter, S. (2001) "Alternate states of ecosystems: evidence and its implications," in N. Huntly and S. Levin (eds) *Ecology: Achievements and Challenges*, London: Blackwell, 357–83.

Cashore, B. (2002) "Legitimacy and the privatization of environmental governance: how Non-State Market-Driven (NSMD) governance systems gain rulemaking authority," *Governance*, 15: 503–29.

Castree, N. (2003) "Commodifying what nature?," *Progress in Human Geography*, 27: 273–97.

—— (2008) "Neoliberalising nature: deregulation and reregulation," *Environment and Planning A*, 40: 131–52.

—— (2010) "Crisis, continuity and change: neoliberalism, the left and the future of capitalism," *Antipode*, 41: 185–213.

Charnowitz, S. (1997) "Two centuries of participation: NGOs and international governance," *Michigan Journal of International Law*, 18: 281–82.

Chisholm, F., Kerry, J., Lane, R., Pattrick, A. and Phillips, G. (2010) "National institutional approaches to climate change," Environmental Governance Series Report 1, School of Environment and Development, University of Manchester, Manchester, UK.

City Repair (2010) "Intersection repair." Online. Available HTTP: http://cityrepair.org/how-to/placemaking/intersectionrepair/ (accessed 29 November 2010).

Clapp, J. and Dauvergne, P. (2005) *Paths to a Green World: The Political Economy of the Global Environment*, London: MIT Press.
Coaffee, J. and Healey, P. (2003) "'My voice, my place': tracking transformations in urban governance," *Urban Studies*, 40: 1979–99.
Coase, R. (1960) "The problem of social cost," *Journal of Law and Economics*, 3: 1–44.
Corbera, E. and Brown, K. (2007) "Building institutions to trade ecosystem services: marketing forest carbon in Mexico," *World Development*, 36: 1956–76.
Cornwall, A. (2004) "New democratic spaces? The politics and dynamics of institutionalized participation," *International Development Studies Bulletin*, 35: 1–10.
Costanza, R. d'Arge, de Groot, R., Farber, S., Grasso, M., Hannon, B., Limburg, K., Naeem, S., O'Neil, R. V., Paruelo, J., Raskin, R. G., Sutton, P. and van den Belt, M. (1997) "The value of the world's ecosystem services and natural capital," *Nature*, 387: 253–60.
Cowan, S. (1998) "Water pollution and abstraction and economic instruments," *Oxford Review of Economic Policy*, 18: 40–49.
Cowell, R. (2003) "Substitution and scalar politics: negotiating environmental compensation in Cardiff Bay," *Geoforum*, 34: 343–58.
Cozijnsen, J., Dudek, D., Meng, K., Petsonk, A. and Eduardo, J. (2007) "CDM and the Post-2012 Framework." Discussion paper, Washington, DC: Environmental Defense.
Crona, B. and Bodin, O. (2006) "What you know is who you know? Communication patterns among resource users as a prerequisite for co-management," *Ecology and Society*, 11. Online. Available HTTP: www.ecologyandsociety.org/vol11/iss2/art7 (accessed 15 January 2011).
Dales, J. (1968) *Pollution, Property and Prices: An Essay in Policy-making and Economics*, New York: Edward Elgar Publishing.
Daly, H. (1991) *Steady-state Economics: Second Edition with New Essays*, Washington, DC: Island Press.
Davies, J. (2002) "The governance of urban regeneration: a critique of the 'governing without government' thesis," *Public Administration*, 80: 301–22.
Davoudi, S. (2006) "The evidence–policy interface in strategic waste planning for urban environments: the 'technical' and 'social' dimensions," *Environment and Planning C*, 24: 681–700.
de Groot, R., Wilson, M. and Boumans, R. (2002) "A typology for the classification, description and valuation of ecosystem functions, goods and services," *Ecological Economics*, 41: 393–408.
De Vivero, J., Mateos, J. and del Corral, D. (2008) "The paradox of public participation in fisheries governance: the rising number of actors and the devolution process," *Marine Policy*, 32: 319–25.
Dean, M. (1999) *Governmentality: Power and Rule in Modern Society*, London: Sage.

Deneven, W. (1992) "The pristine myth: the landscape of the Americas in 1492," *Annals of the Association of American Geographers*, 82: 369–85.

Diamond, P. (1992) "Cosmetic treatment," *Third Way*, 15(6): 18.

Douglas, M. and Wildavsky, A. (1982) *Risk and Culture: Essays on the Selection of Technical and Environmental Dangers*, Berkeley, CA: University of California Press.

Downs, A. (1972) "Up and down with ecology: the issue-attention cycle," *Public Interest*, 28: 38–50.

Drucker, P. (2004) "What makes an effective executive?," *Harvard Business Review*, 82: 58–63.

Dryzek, J. (1997) *The Politics of the Earth: Environmental Discourses*, New York: Oxford University Press.

Duffy, R. (2006) "The potential and pitfalls of global environmental governance: the politics of trans-frontier conservation areas in Southern Africa," *Political Geography*, 25: 89–112.

Dunlap, R. and York, R. (2008) "The globalization of environmental concern and the limits of the postmaterialist values explanation: evidence from multinational surveys," *Sociological Quarterly*, 49: 529–63.

Dupuy, J.-P. (2007) "The catastrophe of Chernobyl twenty years later," *Estudos Avançados*, 21: 243–52.

Easterlin, R. (1974) "Does economic growth improve the human lot? Some empirical evidence," in P. David and M. Reder (eds) *Nations and Households in Economic Growth: Essays in Honor of Moses Abramovitz*, New York: Academic Press.

Eden, S. (2009) "The work of environmental governance networks: traceability, credibility and certification by the Forest Stewardship Council," *Geoforum*, 40: 383–94.

Ellerman, D., Joskow, P., Schmalensee, R., Montero, J. and Bailey, E. (2000) *Markets for Clean Air: The U.S. Acid Rain Program*, New York: Cambridge University Press.

Elmqvist, T. (2008) "Social-ecological systems in transition," *Environmental Sciences*, 5: 69–71.

Enkvist, P., Nauclér, T. and Rosander, J. (2007) "A cost curve for greenhouse gas reduction," *The McKinsey Quarterly*, 1: 34–45.

Environmental Protection Agency (1990) *Clean Air Act*. Online. Available HTTP: www.epa.gov/air/caa/ (accessed 3 August 2009).

Ereaut, G. and Segnit, N. (2006) *Warm Words: How Are We Telling the Climate Story and How Can We Tell It Better?*, London: Institute for Public Policy Research.

Evans, J. (2011) "Adaptation, ecology and the politics of the experimental city," *Transactions of the Institute of British Geographers*, 36: 223–37.

Evans, J., Jones, P. and Krueger, R. (2009) "Organic regeneration and sustainability or can the credit crunch save our cities?," *Local Environment*, 14: 683–98.

Evans, J. and Karvonen, A. (2011) "Living laboratories for sustainability: exploring the politics and epistemology of urban adaptation," in H. Bulkeley, V. Castán Broto, M. Hodson and S. Marvin (eds) *Cities and Low Carbon Transitions*, London: Routledge.

Evernden, N. (1992) *The Social Creation of Nature*, Baltimore, MD: Johns Hopkins University Press.

Fairbrass, J. and Jordan, A. (2005) "Multi-level governance and environmental policy," in I. Bache and M. Flinders (eds) *Multi-level Governance*, Oxford: Oxford University Press, 147–64.

Fairclough, N. (1992) *Discourse and Social Change*, Cambridge: Polity.

Falk, R. (1995) *On Humane Governance: Toward a New Global Politics*, Cambridge: Polity.

Farber, D. (2007) "Basic compensation for victims of climate change," *University of Pennsylvania Law Review*, 155: 651–56.

Fischer, D. and Freudenberg, W. (2001) "Ecological modernization and its critics: assessing the past and looking toward the future," *Society and Natural Resources*, 14: 701–9.

Fischer, F. (2000) *Citizens, Experts and the Environment*, Durham, NC: Duke University Press.

Fischhoff, B. (1995) "Risk perception and communication unplugged: twenty years of process," *Risk Analysis*, 15: 137–45.

Fogel, C. (2004) "The local, the global, and the Kyoto Protocol," in S. Jasanoff and M. Martello (eds) *Earthly Politics: Local and Global in Environmental Governance*, Cambridge, MA: MIT Press.

Folke, C., Carpenter, S., Elmqvist, T., Gunderson, L., Holling, C. and Walker, B. (2002) "Resilience and sustainable development: building adaptive capacity in a world of transformations," *Ambio*, 31: 437–40.

Foucault, M. (1977) *Discipline and Punish*, New York: Pantheon.

—— (1980) *Power/Knowledge: Selected Interviews and Other Writings, 1972–1977*, New York: Pantheon Books.

—— (1991) "Governmentality," in G. Burchall, C. Gordon and P. Miller (eds) *The Foucault Effect: Studies in Governmentality*, Chicago: Chicago University Press.

Friedman, M. (1962) *Capitalism and Freedom*, Chicago: Chicago University Press.

Frosch, R. and Gallopoulos, N. (1989) "Strategies for manufacturing," *Scientific American*, 261: 144–52.

Fues, T., Messner, D. and Scholz, I. (2005) "Global environmental governance from a North–South perspective," in A. Rechkemmer (ed.) *UNEO: Towards an International Environment Organization*, Baden-Baden: Nomos, 241–63.

Funtowicz, S. and Ravetz, J. (1992) "Three types of risk assessment and the emergence of postnormal science," in S. Krimsky and D. Golding (eds) *Social Theories of Risk*, New York: Greenwood Press, 251–73.

Galiana, I. and Green, C. (2009) "Let the global technology race begin," *Nature*, 462: 570–71.

Gamble, A. (1992) *The Free Economy and the Strong State: Politics of Thatcherism*, Cambridge: Polity Press.

Garvey, J. (2008) *The Ethics of Climate Change: Right and Wrong in a Warming World*, London: Continuum.

Geels, F. (2002) "Technological transitions as evolutionary reconfiguration processes: a multilevel perspective and a case study," *Research Policy*, 31: 1257–74.

—— (2004) "From sectoral systems of innovation to socio-technical systems: insights about dynamics and change from sociology and institutional theory," *Research Policy*, 33: 897–920.

Geels, F., Monaghan, A., Eames, M. and Steward, F. (2008) *The Feasibility of Systems Thinking in Sustainable Consumption and Production Policy: A Report to the Department for Environment, Food and Rural Affairs*, London: DEFRA.

Gemmill, B. and Bamidele-Izu, B. (2002) "The role of NGOs and civil society in global environmental governance," in D. Esty and M. Ivanova (eds) *Global Environmental Governance: Options and Opportunities*, New Haven, CT: Yale Center for Environmental Law and Policy, 121–40.

German Advisory Council on Global Change (2009) *Solving the Climate Dilemma: The Budget Approach*, Berlin: German Advisory Council on Global Change.

Giddens, A. (1990) *The Consequences of Modernity*, Cambridge: Polity Press.

—— (2002) *Runaway World: How Globalization is Reshaping Our World*, London: Profile Books.

Gilderbloom, J., Hanka, M. and Lasley, C. (2009) "Amsterdam: planning and policy for the ideal city?," *Local Environment*, 14: 473–93.

Glasbergen, P., Biermann, F. and Mol, A. (eds) (2007) *Partnerships, Governance, and Sustainable Development: Reflections on Theory and Practice*, Cheltenham: Edward Elgar.

Goldemberg, J., Squitieri, R., Stiglitz, J., Amano, A., Shaoxiong, X., Kane, S., Reilly, J. and Teisberg, T. (1996) "Introduction: scope of the assessment," in J. Bruce, H. Lee and E. Haites (eds) *Climate Change 1995: Economic and Social Dimensions of Climate Change. Contribution of Working Group III to the Second Assessment Report of the Intergovernmental Panel on Climate Change*, Cambridge: Cambridge University Press, 17–51.

Gottlieb, B. (2007) *Reinventing Los Angeles: Nature and Community in the Global City*, Cambridge, MA: MIT Press.

Granovetter, M. (1973) "The strength of weak ties," *American Journal of Sociology*, 78: 1360–80.

Grimble, R. and Wellard, K. (1997) "Stakeholder methodologies in natural resource management: a review of concepts, contexts, experiences and opportunities," *Agricultural Systems*, 55: 173–93.

Grubb, M., Vrolijk, C. and Brack, D. (1999) *The Kyoto Protocol: A Guide and Assessment*, London: Royal Institute of International Affairs.

Guha, R. and Martinez-Alier, J. (1997) *Varieties of Environmentalism: Essays North and South*, London: Earthscan.

Gulbrandsen, L. (2010) *Transnational Environmental Governance: The Emergence and Effects of the Certification of Forests and Fisheries*, Cheltenham: Edward Elgar.

Gunderson, L. (1999) "Resilience, flexibility and adaptive management: antidotes for spurious certitude?," *Conservation Ecology*, 3(1): 7.

—— (2000) "Ecological resilience: in theory and application," *Annual Review of Ecology and Systematics*, 31: 425–39.

Gunderson, L. and Holling, C. (eds) (2002) *Panarchy: Understanding Transformations in Human and Natural Systems*, Washington, DC: Island Press.

Gunderson, L., Holling, C. and Light, S. (1995) *Barriers and Bridges to Renewal of Ecosystems and Institutions*, New York: Columbia University Press.

Haas, P. (1990) *Saving the Mediterranean: The Politics of International Environmental Cooperation*, New York: Columbia University Press.

—— (1992) "Banning chlorofluorocarbons: epistemic community efforts to protect the stratospheric ozone," *International Organization*, 46: 187–224.

Habermas, J. (1984) *The Theory of Communicative Action Volume 1: Reason and the Rationalization of Society*, Boston: Beacon Press.

Hajer, M. (2003) "Policy without polity? Policy analysis and the institutional void," *Policy Sciences*, 36: 175–95.

Hajer, M. and Wagenaar, H. (eds) (2003) *Deliberative Policy Analysis: Understanding Governance in the Network Society*, Cambridge: Cambridge University Press.

Hamdi, N. (2004) *Small Change: About the Art of Practice and the Limits of Planning in Cities*, London: Earthscan.

Hamilton, G. (2009) "Public–private partnerships and foreign direct investment as a means of securing a sustainable recovery." Paper delivered at the Fourth Columbia International Investment Conference, Columbia University, 5–6 November.

Hansen, J. (2006) "Our planet's keeper," *New York Review of Books*. Online. Available HTTP: www.nybooks.com/articles/archives/2006/jul/13/the-threat-to-the-planet/?page=2 (accessed 18 November 2010).

Hardin, R. (1968) "The tragedy of the commons," *Science*, 163: 1243–48.

Harrison, C., Burgess, J. and Clark, J. (1998) "Discounted knowledges: farmers' and residents' understandings of nature conservation goals and policies," *Journal of Environmental Management*, 54: 305–20.

Harvey, D. (1996) *Justice, Nature and the Geography of Difference*, Oxford: Blackwell.

—— (2007) *A Brief History of Neoliberalism*, Oxford: Oxford University Press.

Harvey, F. (2007) "Beware the carbon offsetting cowboys," *Financial Times*. Online. Available HTTP: www.ft.com/cms/s/0/dcdefef6-f350-11db-9845-000b5df10621.html#axzz1LZcGMBTN (accessed 12 September 2009).

Hawkins, K. (1984) *Environment and Enforcement: Regulation and the Social Definition of Pollution*, Oxford: Oxford University Press.

Hayek, F. (1948) *Individualism and Economic Order*, Chicago: University of Chicago Press.

Heclo, H. (1974) *Modern Social Politics in Britain and Sweden*, New Haven, CT: Yale University Press.

Heinelt, H. (2007) "Participatory governance and European democracy," in B. Kohler-Kock and B. Rittberger (eds) *Debating the Democratic Legitimacy of the European Union*, London: Rowman & Littlefield, 217–32.

Helm, C. (2000) *Economic Theories of Environmental Cooperation*, Cheltenham: Edward Elgar.

Helm, D., Smale, R. and Phillips, J. (2007) *Too Good to be True?*, London: Vivid Economics Ltd.

Herod, A., O'Tuathail, G. and Roberts, S. (eds) (1998) *An Unruly World? Globalization, Governance and Geography*, London: Routledge.

Hinchliffe, S. (2001) "Indeterminacy indecisions: science, policy and politics in the BSE crisis," *Transactions of the Institute of British Geographers*, 26: 182–204.

Hobbes, T. (1968) (orig. 1651) *Leviathan*, Harmondsworth: Penguin.

Hodson, M. and Marvin, S. (2007) "Understanding the role of the national exemplar in constructing 'strategic glurbanization,'" *International Journal of Urban and Regional Research*, 31: 303–25.

—— (2009) "Cities mediating technological transitions: understanding visions, intermediation and consequences," *Technology Analysis and Strategic Management*, 21: 515–34.

Hodson, M., Marvin, S. and Hewitson, A. (2008) "Constructing a typology of H2 in cities and regions: an international review," *International Journal of Hydrogen Energy*, 33: 1619–29.

Holling, C. (1973) "Resilience and stability of ecological systems," *Annual Review of Ecology and Systematics*, 4: 1–24.

—— (1993) "Investing in research for sustainability," *Ecological Applications*, 3: 552–55.

—— (2004) "From complex regions to complex worlds," *Ecology and Society*, 9: 11. Online. Available HTTP: www.ecologyandsociety.org/vol9/iss1/art11 (accessed 12 December 2009).

Holston, J. (1999) "Spaces of insurgent citizenship," in J. Holston (ed.) *Cities and Citizenship*, London: Duke University Press, 155–73.

Hoogma, R., Kemp, R., Schot, J. and Truffer, B. (2002) "Experimenting for sustainable transport: the findings from HarmoniCOP European case studies," *Environmental Science and Policy*, 8: 287–99.

Hopkins, R. (2008) *The Transition Handbook: From Oil Dependency to Local Resilience*, Totnes: Green.

Houser, T., Mohan, S. and Heilmayr, R. (2009) *A Green Global Recovery? Assessing US Economic Stimulus and the Prospects for International Coordination*, Policy Brief PB09–3, Washington, DC: World Resources Institute.

Hughes, O. (2003) *Public Management and Administration*, Basingstoke: Palgrave Macmillan.

Hulme, M. (2009) *Why We Disagree About Climate Change: Understanding Controversy, Inaction and Opportunity*, Cambridge: Cambridge University Press.

Humphrey, D. (1996) *Forest Politics: The Evolution of International Cooperation*, London: Earthscan.

Hyden, G. (1992) "Governance and the study of politics," in G. Hyden and M. Bratton (eds) *Governance and Politics in Africa*, Boulder, CO: Lynne Rienner.

IPCC (International Panel on Climate Change) (2007) *Fourth Assessment Report*. Online. Available HTTP: www.ipcc.ch/ (accessed 3 April 2009).

Irwin, A. (1995) *Citizen Science*, London: Routledge.

Jackson, T. (2009) *Prosperity Without Growth: Economics for a Finite Planet*, London: Earthscan.

Janssen, M., Schoon, M., Ke, W. and Börner, K. (2006) "Scholarly networks on resilience, vulnerability and adaptation within the human dimensions of global environmental change," *Global Environmental Change*, 16(3): 240–52.

Jasanoff, S. (2004) "Heaven and Earth: images and models of environmental change," in S. Jasanoff and M. Martello (eds) *Earthly Politics: Local and Global in Environmental Governance*, Cambridge, MA: MIT Press, 31–52.

Jasanoff, S. and Wynne, B. (1998) "Is science socially constructed and can it still inform public policy and decision-making?," in S. Rayner and E. Malone (eds) *Human Choice and Climate Change*, Columbus, OH: Battelle Press, 1–88.

Jenkins, M. (2008) "Mother nature's sum," *Miller-McCune*, 1 (October): 44–53.

Jessop, B. (1994) "Postfordism and the state," in A. Amin (ed.) *PostFordism: A Reader*, Oxford: Blackwell, 251–79.

—— (1999) "The dynamics of partnership and governance failure," in G. Stoker (ed.) *The New Management of British Local Governance*, Basingstoke: Macmillan, 11–32.

—— (2003) "Governance and metagovernance: on reflexivity, requisite variety, and requisite irony," in H. Bang (ed.) *Governance, Governmentality and Democracy*, Manchester: Manchester University Press, 142–72.

John, D. (1994) *Civic Environmentalism: Alternatives to Regulation in States and Communities*, Washington, DC: CQ Press.

John, P. and Cole, A. (2000) "Policy networks and local political leadership in Britain and France," in G. Stoker (ed.) *The New Politics of British Local Governance*, London: Palgrave Macmillan, 72–90.

Jones, C., Hesterly, W. and Borgatti, S. (1997) "A general theory of network governance: exchange conditions and social mechanisms," *Academy of Management Review*, 22: 911–45.

Jones, P. and Evans, J. (2006) "Urban regeneration and the state: exploring notions of distance and proximity," *Urban Studies*, 43: 1491–1509.

—— (2008) *Urban Regeneration in the UK: Theory and Practice*, London: Sage.

Joosten, H. and Couwenberg, J. (2008) "Peatlands and carbon," in F. Parish, A. Sirin, D. Charman, H. Joosten, T. Minayeva, M. Silvius and L. Stringer (eds) *Assessment on Peatlands, Biodiversity and Climate Change: Main Report*, Wageningen, the Netherlands: Global Environment Centre, Kuala Lumpur and Wetlands International, 155–79.

Jordan, A. (2002) *Environmental Policy in the European Union: Actors, Institutions and Processes*, London: Earthscan.

Jordan, A. and Jeppesen, T. (2000) "EU environmental policy: adapting to the principle of subsidiarity?," *European Environment*, 10(2): 64–74.

Jordan, A. and O'Riordan, T. (2003) "Institutions for global environmental change," *Global Environmental Change*, 13: 223–28.

Jordan, A. and Voisey, H. (1998) "The 'Rio Process': the politics and substantive outcomes of 'Earth Summit II,'" *Global Environmental Change*, 8: 93–97.

Jordan, A., Wurzel, R. K. W. and Zito, A. R. (2003) "Comparative conclusions. 'New' environmental policy instruments: an evolution or a revolution in environmental policy?," *Environmental Politics*, 12(1): 201–24.

Karvonen, A. and Yocum, K. (2011) "The civics of urban nature: enacting hybrid landscapes," *Environment and Planning A*, 43(6): 1305–22.

Kates, R., Clark, W., Corell, R., Hall, J., Jaeger, C. C., Lowe, I., McCarthy, J., Schellhuber, H., Bolin, B., Dickson, N., Faucheux, S., Gallopin, G., Grubler, A., Huntley, B., Jäger, J., Jodha, N., Kasperson, R., Mabogunje, A., Matson, P., Mooney, H., More, III B., O'Riordan, T. and Svedin, U. (2001) "Sustainability science," *Science*, 292: 641–42.

Kemp, R., Parto, S. and Gibson, R. (2005) "Governance for sustainable development: moving from theory to practice," *International Journal of Sustainable Development*, 8: 12–30.

Kemp, R., Rip, A. and Schot, J. (2001) "Constructing transition paths through the management of niches," in R. Garud and P. Karnoe (eds) *Path Dependence and Creation*, Mahwah, NJ: Lawrence Erlbaum Associates, 269–99.

Kemp, R., Rotmans, J. and Loorbach, D. (2007) "Assessing the Dutch energy transition policy: how does it deal with managing dilemmas of managing transition?," *Journal of Environment Policy and Planning*, 9: 315–31.

Kemp, R., Schot, J. and Hoogma, R. (2001a) "Regime shifts to sustainability through processes of niche formation: the approach of strategic niche management," *Technology Analysis and Strategic Management*, 10: 175–96.

Keohane, R. and Nye, J. (eds) (1971) *Transnational Relations and World Politics*. Online. Available HTTP: http://148.201.96.14/dc/ver.aspx?ns=000193531 (accessed 2 May 2010).

Kersbergen, K. and Waarden, F. (2004) "'Governance' as a bridge between disciplines: cross-disciplinary inspiration regarding shifts in governance and problems of governability, accountability and legitimacy," *European Journal of Political Research*, 43: 143–71.

Kickert, W., Klijn, E. and Koppenjan, J. (1999) *Managing Complex Networks: Strategies for the Public Sector*, London: Sage.

Kirsch, D. (2000) *The Electric Vehicle and the Burden of History*, London: Rutgers University Press.

Kjaer, A. (2008) *Governance*, Cambridge: Polity Press.

Klein, N. (2007) *The Shock Doctrine: The Rise of Disaster Capitalism*, New York: Metropolitan Books/Henry Holt.

Klein, R., Schipper, L. and Dessai, S. (2003) "Integrating mitigation and adaptation into climate and development policy: three research questions," Working Paper 40, Norwich: Tyndall Centre for Climate Change Research.

Klijn, E. and Skelcher, C. (2007) "Democracy and governance networks: compatible or not?," *Public Administration*, 85: 587–608.

Kooiman, J. (ed.) (1993) *Modern Governance: New Government–Society Interactions*, London: Sage.

—— (1999) "Social-political governance," *Public Management Review*, 1: 67–92.

—— (2000) "Societal governance: levels, models and orders of social political interaction," in J. Pierre (ed.) *Debating Governance: Authority, Steering and Democracy*, Oxford: Oxford University Press, 138–66.

—— (2003) *Governing as Governance*, London: Sage.

Koontz, T. (2003) "An introduction to the institutional analysis and development framework for forest management research." Paper prepared for the "First Nations and Sustainable Forestry: Institutional Conditions for Success" workshop, University of British Columbia, Vancouver.

Krasner, S. (1983) "Structural causes and regime consequences: regimes as intervening variables," in S. Krasner (ed.) *International Regimes*, Ithaca, NY: Cornell University Press.

Krueger, R. and Savage, L. (2007) "City-regions and social reproduction: a 'place' for sustainable development?," *International Journal of Urban and Regional Research*, 31: 215–23.

Kutting, G. and Lipschutz, R. (2009) *Environmental Governance: Power and Knowledge in a Local-Global World*, London: Routledge.

Kwa, C. (1987) "Representations of nature mediating between ecology and science policy: the case of the International Biological Program," *Social Studies of Science*, 17: 413–42.

Landy, M., Roberts, M. and Thomas, S. (1994) *The Environmental Protection Agency: Asking the Wrong Questions, From Nixon to Clinton, Expanded Edition*, New York: Oxford University Press.

Landy, M. and Rubin, C. (2001) *Civic Environmentalism: A New Approach to Policy*, Washington, DC: George Marshall Institute.

Latour, B. (1993) *We Have Never Been Modern*, London: Harvester Wheatsheaf.

Leichenko, R., O'Brien, K. and Solecki, W. (2010) "Climate change and the global financial crisis: a case of double exposure," *Annals of the Association of American Geographers*, 100(4): 963–72.

Lenton, T., Held, H., Kriegler, E., Hall, J., Lucht, W., Rahmstork, S. and Joachim Schellnhuber, H. (2008) "Tipping elements in the Earth's climate system," *Proceedings of the National Academy of Sciences*, 105: 1786–93.

Lessig, L. (2001) *The Future of Ideas: The Fate of the Commons in a Connected World*, New York: Random House.

Levin, K., Cashore, B., Bernstein, S. and Auld, G. (forthcoming) *Playing It Forward: Path Dependency, Progressive Incrementalism, and the "Super Wicked" Problem of Global Climate Change.*

Levin, S. (1998) "Ecosystems and the biosphere as complex adaptive systems," *Ecosystems*, 1: 431–36.

Levin, S., Barrett, S., Aniyar, S., Baumol, W., Bliss, C., Bolin, B., Dasgupta, P., Ehrlich, P., Folke, C., Gren, I., Holling, C., Jansson, A., Jansson, B., Mäler, K., Martin, D., Perrings, C. and Sheshinski, E. (1998) "Resilience in natural and socio-economic systems," *Environment and Development Economics*, 3: 222–35.

Lindblom, C. (1979) "Still muddling, not yet through," *Public Administration Review* (November/December): 517–26.

Lipschutz, R. (1996) *Global Civil Society and Global Environmental Governance: The Politics of Nature from Place to Planet*, New York: State University of New York Press.

Litfin, K. (1994) *Ozone Discourses: Science and Politics in Global Environmental Cooperation*, New York: Columbia University Press.

Lodefalk, M. and Whalley, J. (2002) "Reviewing proposals for a world environmental organisation," *The World Economy*, 25(5): 601–17.

Lohmann, L. (2006), "Carbon trading: a critical conversation on climate change, privatisation and power," *Development Dialogue*, 48 (September): 73.

Lomborg, B. (ed.) (2007) *Smart Solutions to Climate Change: Comparing Costs and Benefits*, Cambridge: Cambridge University Press.

Lowe, P. and Ward, S. (eds) (1998) *British Environmental Policy and Europe: Politics and Policy in Transition*, London: Routledge.

Lowndes, V. (1996) "Varieties of new institutionalism: a critical appraisal," *Public Administration*, 74: 181–97.

—— (2001) "Rescuing Aunt Sally: taking institutional theory seriously in urban politics," *Urban Studies*, 38: 1953–71.

Lowndes, V. and Skelcher, C. (1998) "The dynamics of multiorganizational partnerships: an analysis of changing modes of governance," *Public Administration*, 76: 313–33.

Luke, T. (1994) "Worldwatching as the limits to growth," *Capitalism, Nature, Socialism*, 5: 43–64.

—— (1999) "Environmentality as governmentality," in E. Darier (ed.) *Discourses of the Environment*, Oxford: Blackwell, 121–51.

Lytle, M. (2007) *The Gentle Subversive: Rachel Carson,* Silent Spring, *and the Rise of the Environmental Movement*, New York: Oxford University Press.

Macchiavelli, N. (1992) *The Prince*, London: W. W. Norton.

Malthus, T. (1970) (orig. 1798) *An Essay on the Principal of Population*, London: Penguin Books.

Mansfield, B. (2006) "Assessing market-based environmental policy using a case study of North Pacific fisheries," *Global Environmental Change*, 16: 29–39.

March, J. and Olsen, J. (1984) "The new institutionalism: organizational factors in political life," *The American Political Science Review*, 74: 734–49.

Masdar City (2010) Masdar City Website. Available HTTP: www.masdarcity.ae (accessed 21 March 2010).

McCormick, J. (1991) *British Politics and the Environment*, London: Earthscan.

—— (2005) "The role of environmental NGOs in international regimes," in N. Vig and R. Axelrod (eds) *The Global Environment: Institutions, Law and Policy*, London: Earthscan, 52–71.

McIlgorm, A., Hanna, S., Knapp, G., Le Floc'H, P., Milled, F. and Pan, M. (2010) "How will climate change alter fishery governance? Insights from seven international case studies," *Journal of Marine Policy*, 32: 170–77.

McKibben, B. (2007) *Deep Economy: The Wealth of Communities and the Durable Future*, New York: Times Books.

McKinsey and Company (2009) "Pathways to a low carbon economy: version 2 of the global abatement cost-curve." Online. Available HTTP: www.mckinsey.com/globalGHGcostcurve (accessed 27 November 2010).

Meadowcroft, J. (2009) "What about the politics? Sustainable development, transition management, and long term energy transitions," *Policy Science*, 42: 323–40.

Meadows, D. (1972) *The Limits to Growth: A Report for the Club of Rome Project on the Predicament of Mankind*, New York: Universe Books.

Meinshausen, M. (2006) "What does a 2°C target mean for greenhouse gas concentrations? A brief analysis based on multi-gas emission pathways and several climate sensitivity uncertainty estimates," in H. Schellnhuber, W. Cramer, N. Nakicenovic, T. Wigley and G. Yohe (eds) *Avoiding Dangerous Climate Change*, Cambridge: Cambridge University Press, 265–79.

Mitchell, R. (2010) *International Environmental Agreements Database Project (Version 2010.2)*. Online. Available HTTP: http://iea.uoregon.edu/ (accessed 8 September 2010).

Mol, A. (1995) *The Refinement of Production: Ecological Modernisation Theory and the Chemical Industry*, Utrecht, the Netherlands: van Arkel.

Moneva, J. and Archel, J. (2006) "GRI and the camouflaging of corporate unsustainability," *Accounting Forum*, 30: 121–37.

Moody-Stuart, M. (2008) *Society Depends on More for Less*. Online. Available HTTP: http://news.bbc.co.uk/1/hi/sci/tech/7218002 (accessed 14 September 2010).

Morley, D. and Robins, K. (1995) *Spaces of Identity: Global Media, Electronic Landscapes and Cultural Boundaries*, London: Routledge.

Myers, N. and Golubiewski, N. (2007) "Perverse subsidies," in C. Cleveland (ed.) *Encyclopedia of Earth*, Washington, DC: Environmental Information Coalition, National Council for Science and the Environment.

Myers, N. and Kent, J. (2001) *Perverse Subsidies: Tax $s Undercutting Our Economies and Environments Alike*, Washington, DC: Island Press.

Najam, A. (2003) "The case against a new international environmental organization," *Global Governance*, 9(3): 367–84.

Najam, A., Huq, S. and Sokona, Y. (2003) "Climate negotiations beyond Kyoto: developing countries' concerns and interests," *Climate Policy*, 3: 221–31.

Nash, L. (2006) *Inescapable Ecologies*, Los Angeles: University of California Press.

Newman, L. and Dale, A. (2005) "Network structure, diversity, and proactive resilience building: a response to Tompkins and Adger," *Ecology and Society*, 10. Online. Available HTTP: www.ecologyandsociety.org/vol10/iss1/resp2 (accessed 15 February 2011).

Nordquist, J. (2006) *Evaluation of Japan's Top Runner Programme*. Online. Available HTTP: www.aidee.org/ (accessed 17 September 2010).

North, D. and Weingast, B. (1989) "Constitutions and commitment: the evolution of institutional governing public choice in seventeenth century England," *The Journal of Economic History*, 49(4): 803–32.

Oberthür, S. and Gehring, T. (2004) "Reforming international environmental governance: an institutionalist critique of the proposal for a World Environmental Organisation," *Politics, Law and Economics*, 4: 359–81.

Oberthür, S. and Ott, H. (1999) *The Kyoto Protocol: International Climate Policy for the 21st Century*, Berlin: Springer Verlag.

OECD (2010) *International Development Statistics* (IDS) online databases on aid and other resource flows. Online. Available HTTP: http://blds.ids.ac.uk/elibrary/db_stats.html (accessed 22 November 2010).

O'Neill, B. and Oppenheimer, M. (2002) "Climate change: dangerous climate impacts and the Kyoto Protocol," *Science*, 296: 55–75.

O'Neill, J. (2007) *Markets, Deliberation, and Environment*, New York: Routledge.

Ostrom, E. (1990) *Governing the Commons: The Evolution of Institutions for Collective Action*, Cambridge: Cambridge University Press.

Ostrom, E., Gardner, R. and Walker, J. (1994) *Rules, Games, and Common Pool Resources*, Ann Arbor: University of Michigan Press.

Owens, S. and Driffil, L. (2008) "How to change attitudes and behaviours in the context of energy," *Energy Policy*, 36: 4412–18.

Pahl-Wostl, C. (2007) "Transition towards adaptive management of water facing climate and global change," *Water Resources Management*, 21(1): 49–62.

Park, J., Conca, K. and Finger, M. (eds) (2008) *The Crisis of Global Environmental Governance*, London: Routledge.

Parker, C., Mitchell, A., Trivedi, M. and Mardas, M. (2008) *The Little REDD Book: A Guide to Governmental and Non-governmental Proposals for Reducing Emissions from Deforestation and Degradation*, Oxford: Global Canopy Foundation. Online. Available HTTP: www.amazonconservation.org/pdf/redd_the_little_redd_book_dec_08.pdf (accessed 17 April 2011).

Parreno, J. (2007) "Does the current Clean Development Mechanism (CDM) deliver its sustainable development claim? An analysis of officially registered CDM projects," *Climatic Change*, 84(1): 75–90.

Pattberg, P. and Stripple, J. (2008) "Beyond the public and private divide: remapping transnational climate governance in the 21st century," *International Environmental Agreements: Politics, Law and Economics*, 8: 367–88.

Perkins, R. (2003) "Environmental leapfrogging in developing countries: a critical assessment and reconstruction," *Natural Resources Forum*, 27: 177–88.

Perrings, C. (1998) "Resilience in the dynamics of economy–environment systems," *Environmental and Resource Economics*, 11: 503–20.

Petts, J. (1995) "Waste management strategy development: a case study of community involvement and consensus-building in Hampshire," *Journal of Environmental Planning and Management*, 38: 519–36.

—— (ed.) (1999) *Handbook of Environmental Impact Assessment. Volume 1, Environmental Impact Assessment: Process, Methods and Potential*, Oxford: Blackwell.

—— (2006) "Managing public engagement to optimize learning: reflections from urban rivers," *Human Ecology Review*, 13: 172–81.

Pielke, R. Jr (2005) "Misdefining 'climate change': consequences for science and action," *Environmental Science and Policy*, 8: 548–61.

Pierre, J. (2000) "Introduction: understanding governance," in J. Pierre (ed.) *Debating Governance*, Oxford: Oxford University Press, 112.

Pierre, J. and Peters, G. (2000) *Governance, Politics and the State*, London: Macmillan Press.

Pierre, J. and Stoker, G. (2002) "Toward multi-level governance," in P. Dunleavy, A. Gamble, R. Heffernan, I. Holliday and G. Peele (eds) *Developments in British Politics 6*, Basingstoke: Palgrave.

Pierson, P. and Skocpol, T. (2002) "Historical institutionalism in contemporary political science," in I. Katznelson and H. Miller (eds) *Political Science: State of the Discipline*, New York: Norton, 693–721.

Pincetl, S. (2010) "From the sanitary city to the sustainable city: challenges to institutionalizing biogenic (nature's services) infrastructure," *Local Environment*, 15: 43–58.

Polanyi, K. (1944) *The Great Transformation: The Political and Economic Origins of Our Time*, Boston: Beacon Press.

Pollin, R., Garrett-Peltier, H., Heintz, J. and Scharber, H. (2008) *Green Recovery: A Program to Create Good Jobs and Start Building a Low-carbon Economy*, Washington, DC: Center for American Progress.

Powell, W. (1991) "Neither market nor hierarchy: network forms of organisation," in G. Thompson, J. Frances, R. Levacic and J. Mitchell (eds) *Markets, Hierarchies and Networks: The Coordination of Social Life*, London: Sage.

Prell, C., Hubacek, K., Quinn, C. and Reed, M. (2009) "'Who's in the network?' When stakeholders influence data analysis," *Systemic Practice and Action Research*, 21: 443–58.

Prell, C., Hubacek, K. and Reed, M. (2007) *Stakeholder Analysis and Social Network Analysis in Natural Resource Management*, Sustainability Research Institute Papers 6, Leeds: University of Leeds.

Princen, T., Finger, M. and Manno, J. (1994) "Translational linkages," in T. Princen and M. Finger (eds) *Environmental NGOs in World Politics*, London: Routledge, 217–36.

Provan, K. G. and Kenis, P. (2008) "Modes of network governance: structure, management, and effectiveness," *Journal of Public Administration Research and Theory*, 12(2): 229–52.

Pucher, J. (2007) "Case studies of cycling in Amsterdam, The Netherlands," New Brunswick, NJ: Rutgers University, Center for Urban and Economic Research, Working Paper.

Ramus, C. and Montiel, I. (2005) "When are corporate environmental policies a form of greenwashing?," *Business and Society*, 44(4): 377–414.

Rancière, J. (2007) *On the Shores of Politics* (translated by Liz Heron), London: Verso.

Raudsepp-Hearne, C., Peterson, G., Tengö, M., Bennett, E., Holland, T., Benessaiah, K., Macdonald, G. and Pfeifer, L. (2010) "Untangling the environmentalists' paradox: why is human well-being increasing as ecosystem services degrade?" *Bioscience*, 60(8): 576–89.

Redman, C. L. and Kinzig, A. P. (2003) "Resilience of past landscapes: resilience theory, society, and the *longue durée*," *Conservation Ecology*, 7(1): 14. Online. Available HTTP: www.consecol.org/vol7/iss1/art14 (accessed 12 October 2009).

Redman, C., Morgan Grove, J. and Kuby, L. (2004) "Integrating social science into the Long Term Ecological Research (LTER) network: social dimensions of ecological change and ecological dimensions of social change," *Ecosystems*, 7(2): 161–71.

REN21 (2010) *About REN21*. Online. Available HTTP: www.ren21.net/ren21/default.asp (accessed 8 May 2010).

Renn, O. (1999) "A model for an analytic deliberative process in risk management." *Environmental Science and Techology*, 33: 3049–55.

Renn, O., Webler, T. and Wiedemann, P. (eds) (1995) *Fairness and Competence in Citizen Participation: Evaluating Models for Environmental Discourse*, Dordrecht, the Netherlands: Kluwer.

Rhodes, R. (1996) "The new governance: governing without government," *Political Studies*, 44(4): 652–67.

—— (ed.) (1997) *Understanding Governance: Policy Networks, Governance, Reflexivity and Accountability*, Buckingham: Open University Press.

Rhodes, R. and Marsh, D. (1992) "Policy networks in British politics," in D. Marsh and R. Rhodes (eds) *Policy Networks in British Government*, Oxford: Clarendon Press.

Rice, J. (2010) "Climate, carbon and territory: greenhouse gas mitigation in Seattle, Washington," *Annals of the Association of American Geographers*, 100(4): 929–37.

Rip, A. and Kemp, R. (1998) "Technological Change," in S. Raynor and E. Malone (eds) *Human Choice and Climate Change, Vol. 2.*, Columbus, OH: Batelle Press, 327–99.

Risse-Kappen, T. (1995) *Bringing Transnational Relations Back In: Non-state Actors, Domestic Structures, and International Institutions*, Cambridge: Cambridge University Press.

Rittel, H. and Webber, M. (1973) "Dilemmas in a general theory of planning," *Policy Sciences*, 4: 155–69.

Robertson, M. (2004) "The neoliberalization of ecosystem services: wetland mitigation banking and problems in environmental governance," *Geoforum*, 35(3): 361–73.

Rose, A. (1998) "Viewpoint. Global warming policy: who decides what is fair?," *Energy Policy*, 26(1): 1–3.

Rosenau, J. (1995) "Governance in the twenty-first century," *Global Governance*, 1: 13–43.

Rotmans, J., Kemp, R. and van Asselt, M. (2001) "More evolution than revolution: transition management in public policy," *Foresight*, 3(1): 15–31.

Royal Society (2009) *Geoengineering the Climate: Science, Governance and Uncertainty*, London: The Royal Society.

Ruggie, J. (2004) "Reconstituting the global public domain: issues, actors, and practices," *European Journal of International Relations*, 10(4): 499–531.

Rutherford, P. (1999) "The entry of life into history," in E. Darier (ed.) *Discourses of the Environment*, Oxford: Blackwell, 37–62.

Rutherford, S. (2007) "Green governmentality: insights and opportunities in the study of nature's rule," *Progress in Human Geography*, 31(3): 291–307.

Rydin, Y. (2007) "Indicators as a governmentality technology? The lessons of community-based sustainability indicator projects," *Environment and Planning D*, 25(4): 610–24.

—— (2010) *Governing for Sustainable Urban Development*, London: Earthscan.

Sachs, W. (1999) *Planet Dialectics: Explorations in Environment and Development*, New York: Zed Books.

Sagoff, M. (2004) *Price, Principle, and the Environment*, Cambridge: Cambridge University Press.

Saunier, R. and Meganck, R. (2009) *Dictionary and Introduction to Global Environmental Governance*, London: Earthscan.

Sayre, N. (2005) "Ecological and geographical scale: parallels and potential for integration," *Progress in Human Geography*, 29: 276–90.

Scheffer, M., Westley, F. and Brock, W. (2002) "Dynamic interaction of societies and ecosystems: linking theories from ecology, economy, and sociology," in L. Gunderson and C. Holling (eds) *Panarchy: Understanding Transformations in Human and Natural Systems*, Washington, DC: Island Press, 195–240.

Schlager, E. (1999) "A comparison of frameworks, theories, and models of policy processes," in P. Sabatier (ed.) *Theories of the Policy Process*, Boulder, CO: Westview Press.

Schmitter, P. (2002) "Participation in governance arrangements: is there any reason to expect it will achieve 'sustainable and innovative policies in a multi-level context'?," in G. Jürgen and B. Gbikpi (eds) *Participatory Governance: Political and Societal Implications*, Opladen, Germany: Leske & Budrich, 51–70.

Sen, A. (1992) *Inequality Reexamined*, Cambridge, MA: Harvard University Press.

Sepkoski, J. (1997) "Biodiversity: past, present and future," *Journal of Paleontology*, 71: 533–39.

Seyfang, G. (2003) "Environmental megaconferences: from Stockholm to Johannesburg and beyond global," *Environmental Change*, 13: 223–28.

—— (2006) "Ecological citizenship and sustainable consumption: examining local organic food networks," *Journal of Rural Studies*, 22: 383–95.

Seyfang, G. and Jordan, A. (2002) "'Mega' environmental conferences: vehicles for effective, long term environmental planning?," in S. Stokke and O. Thommesen (eds) *Yearbook of International Cooperation on Environment and Development*, London: Earthscan, 19–26.

Shackley, S., Young, P., Parkinson, S. and Wynne, B. (1998) "Uncertainty, complexity, and concepts of good science in climate change modeling: are GCMs the best tools?," *Climatic Change*, 38: 159–205.

Shogren, J. (1998) "Into the wilderness within," *Environment and Development Economics*, 3(2): 221–62.

Shove, E. (2003) *Comfort, Cleanliness and Convenience: The Social Organisation of Normality*, Oxford: Berg.

Simmons, P. (1998) "Learning to live with NGOs," *Foreign Policy*, Fall: 82–96. Online. Availible HTTP: www.globalpolicy.org/component/content/article/177/31607.html (accessed 18 May 2010).

Smil, V. (2002) "Nitrogen and food production: proteins for human diets," *Ambio*, 31: 126–31.

Smith, A. (2010) "Community-led urban transitions and resilience: performing transition towns in a city," in H. Bulkeley, V. Castán Broto, M. Hodson and S. Marvin (eds) *Cities and Low Carbon Transitions*, London: Routledge.

Sneddon, C. (2002) "Water conflicts and river basins: the contradictions of co-management and scale in Northeast Thailand," *Society and Natural Resources*, 15(8): 725–41.

Solow, R. (1974) "The economics of resources or the resources of economics: Richard T. Ely Lecture," *American Economic Review*, May: 1–14.

Sorensen, E. and Torfing, J. (eds) (2007) *Theories of Democratic Governance*, Basingstoke: Palgrave Macmillan.

Spaargaren, G. (1997) *The Ecological Modernisation of Production and Consumption. Essays in Environmental Sociology*, Wageningen, the Netherlands: Wageningen University.

Speth, J. and Haas, P. (2006) *Global Environmental Governance*, Washington, DC: Island Press.

Stern, N. (2009) *A Blueprint for a Safer Planet: How to Manage Climate Change and Create a New Era of Progress and Prosperity*, London: Bodley Head.

Stern, N., Peters, S., Bakhshi, V., Bowen, A., Cameron, C., Catovsky, S., Crane, D., Cruickshank, S., Dietz, S. and Edmonson, N. (2006) *Stern Review: The Economics of Climate Change*, London: HM Treasury.

Stern, P. and Fineberg, H. (1996) *Understanding Risk: Informing Decisions in a Democratic Society*, Washington, DC: National Academy Press.

Steward, F. (2008) *Breaking the Boundaries: Transformative Innovation for the Global Good*, London: National Endowment for Science, Technology and the Arts.

Stirling, A. (1998) "Risk at a turning point?," *Journal of Risk Research*, 1(2): 97–109.

Stoker, G. (1998) "Governance as theory: five propositions," *International Social Science Journal*, 50(155): 17–28.

Stokes, S., Friscia, T. and O'Marah, K. (2008) "Turning the White House into a greenhouse," Alert Article, *AMR Research*. Online. Available HTTP: www.amr.com (accessed 3 March 2009).

Stroup, R. (2003) *Eco-Nomics: What Everyone Should Know About Economics and the Environment*, Washington, DC: Cato Institute.

Sustainable Development Commission (2009) *Prosperity Without Growth? The Transition to a Sustainable Economy*. Online. Available HTTP: www.sd-commission.org.uk/publications.php?id=914 (accessed 3 March 2009).

Swyngedouw, E. (2007) "Impossible sustainability and the post-political condition," in R. Krueger and D. Gibbs (eds) *The Sustainable Development Paradox: Urban Political Economy in the US and Europe*, New York: Guilford Press, 13–40.

Taylor, M. (2007) "Community participation in the real world: opportunities and pitfalls in new governance spaces," *Urban Studies*, 44: 297–317.

Taylor, P. and Buttel, F. (1992) "How do we know we have Global Environmental Problems? Science and the globalization of environmental discourse," *Geoforum*, 23(3): 405–16.

Thomas, D. and Middleton, N. (1994) *Desertification: Exploding the Myth*, Chichester: Wiley.

Thornes, J. and Randalls, S. (2007) "Commodifying the atmosphere: pennies from heaven?," *Physical Geography*, 89(4): 273–85.

Tickell, O. (2008) *Kyoto 2: How to Manage the Global Greenhouse*, London: Zed Books.

Tienhaara, K. (2009) *The Expropriation of Environmental Governance: Protecting Foreign Investors at the Expense of Public Policy*, Cambridge: Cambridge University Press.

Tippett, J., Searle, B., Pahl-Wosti, C. and Rees, Y. (2005) "Social learning in public participation in river basin management: early findings from HarmoniCOP European case studies," *Environmental Science and Policy*, 8(3): 287–99.

Uggla, Y. (2008) "What is this thing called 'natural'? The nature–culture divide in climate change and biodiversity policy," *Journal of Political Ecology*, 17: 79–91.

UNEP, SustainAbility, and Standard & Poor (2004) *Risks and Opportunities; Best Practice in Nonfinancial Reporting*. Online. Available HTTP: http://hqweb.unep.org/Documents.Multilingual/Default.asp?DocumentID =412&ArticleID=4653&l=en (accessed 7 November 2010).

UNFCCC (2001) "Decision 17/CP.7 Modalities and procedures for a Clean Development Mechanism as defined in Article 12 of the Kyoto Protocol." Online. Available HTTP: http://unfccc.int/resource/docs/cop7/13a02.pdf (accessed 15 January 2011).

United Nations (1992) *Earth Summit Agenda 21: The United Nations Programme of Action from Rio*, New York: United Nations Department of Public Information.

—— (2002) *Report of the World Summit on Sustainable Development, Johannesburg*, A/CONF.199/20, New York: United Nations.

—— (2004) *Global Compact*. Online. Available HTTP: www.unglobalcompact.org/ (accessed 2 October 2009).

United States Geological Survey (2007) *Sea Level Rise*. Online. Available HTTP: http://cegis.usgs.gov/sea_level_rise.html (accessed 24 November 2010).

Unruh, G. and Carillo-Hermosilla, J. (2006) "Globalizing carbon lock-in," *Energy Policy*, 34(10): 1185–97.

Uzzi, B. (1997) "Social structure and competition in interfirm networks: the paradox of embeddedness," *Administrative Science Quarterly*, 42: 35–67.

Verheyen, R. (2002) "Adaptation to the impacts of anthropogenic climate change: the international legal framework," *Review of European Community and International Environmental Law*, 11(2): 129–43.

Vis, M., Klijn, F., de Bruijn, K. and Buuren, M. (2003) "Resilience strategies for flood risk management in the Netherlands," *International Journal of River Basin Management*, 1: 33–40.

Vogel, D. (2006) *The Market for Virtue: The Potential and Limits of Corporate Social Responsibility*, Washington, DC: Brookings Institution Press.

Walker, B., Carpenter, S., Anderies, J., Abel, N., Cumming, G., Janssen, M., Lebel, L., Norberg, J., Peterson, G. and Pritchard, R. (2002) "Resilience management in social ecological systems: a working hypothesis for a participatory approach," *Conservation Ecology*, 6(1): 14.

Walker, B., Holling, C. S., Carpenter, S. R. and Kinzig, A. (2004) "Resilience, adaptability and transformability in social–ecological systems," *Ecology and Society*, 9(2): 5.

Walker, J. (2009) "The strange evolution of Holling's resilience, or the resilience of economics and the eternal return of infinite growth." Paper presented at "Cities, Nature. Justice," University of Technology, Sydney, February 2009.

Wang, T. and Watson, J. (2007) "Who owns China's carbon emissions?," Tyndall Briefing Note no. 23. Online. Available HTTP: http://tyndall.webapp1.uea.ac.uk/publications/briefing_notes/bn23.pdf (accessed 1 April 2008).

Warren, R., (2006) "Impacts of global climate change at different annual mean global temperature increases," in H. Schellnhuber, W. Cramer, N. Nakicenovic, T. Wigley and G. Yohe (eds) *Avoiding Dangerous Climate Change*, Cambridge: Cambridge University Press, 93–131.

Weart, S. (2008) *The Discovery of Global Warming*, Cambridge, MA: Harvard University Press.

Weather Risk Management Association (2010) *Trading Weather Risk*. Online. Available HTTP: www.wrma.org/risk_trading.html (accessed 12 March 2011).

Weber, N. and Christopherson, T. (2002) "The influence of nongovernmental organisations on the creation of Natura 2000 during the European Policy Process," *Forest Policy and Economics*, 4(1): 1–12.

Webler, T., Kastenholz, H. and Renn, O. (1995) "Public participation in impact assessment: a social learning perspective," *Environmental Impact Assessment Review*, 15: 443–63.

Webster, K. and Johnson, C. (2008) *Sense and Sustainability: Educating for a Low Carbon World*, Yorkshire: TerraPreta.

Weisman, A. (2007) *The World Without Us*, New York: St Martin's Press.

Welford, R. and Starkey, R. (1996) *The Earthscan Reader in Business and the Environment*, London: Earthscan.

White, I. and Howe, J. (2003) "Planning and the European Union Water Framework Directive," *Journal of Environmental Planning and Management*, 46: 62–131.

Whitehead, A. (1948) *Science and the Modern World*, New York: Mentor.

Willetts, P. (2002) *What is a "Nongovernmental Organization?" Output from the Research Project on Civil Society Networks in Global Governance.* Online. Available HTTP: www.staff.city.ac.uk/p.willetts/index.htm (accessed 28 October 2010).

World Bank (2004) *Corporate Social Responsibility and Labor*. Online. Available HTTP: www.worldbank.org/privatesector (accessed 5 December 2010).

World Commission on Environment and Development (1987) *Brundtland Report: Our Common Future*, Oxford: Oxford University Press.

Worster, D. (1977) *Nature's Economy: A History of Ecological Ideas*, New York: Sierra Club Books.

Wynne, B. (1996) "May the sheep safely graze? A reflexive view of the expert/lay knowledge divide," in S. Lash, B. Szerszynski and B. Wynne (eds) *Risk, Environment and Modernity: Towards a New Ecology*, London: Sage, 44–83.

Young, O. (1982) *Resource Regimes: Natural Resources and Social Institutions*, Berkeley: University of California Press.

—— (2008) "The architecture of global environmental governance: bringing science to bear on policy," *Global Environmental Politics*, 8(1): 14–32.

Young, S. (2000) "The origins and evolving nature of ecological modernisation," in S. Young (ed.) *The Emergence of Ecological Modernisation*, London: Routledge, 1–40.

Zizek, S. (2008) *In Defence of Lost Causes*, London: Verso.

Index

Page numbers in *Italics* represent tables.
Page numbers in **Bold** represent figures.
Page numbers followed by b represent box.